P9-AON-572

THE UNFINISHED UNIVERSE

Louise B. Young

SIMON AND SCHUSTER
New York

Copyright © 1986 by Louise B. Young. All rights reserved,
including the right of reproduction in whole or in part in
any form. Published by Simon and Schuster, a division of
Simon & Schuster, Inc., Simon & Schuster Building, Rocke-
feller Center, 1230 Avenue of the Americas, New York,
New York 10020. SIMON AND SCHUSTER and colophon
are registered trademarks of Simon & Schuster, Inc. Designed by Irving Perkins Associates. Manufactured in the
United States of America.

10 9 8 7 6 5 4 3 2 1

Library of Congress Cataloging in Publication Data
Young, Louise B.
The unfinished universe.
Includes index.
1. Science. 2. Cosmology. 3. Evolution. I. Title.
Q158.5.Y68 1986   500   85-22165

ISBN: 0-671-52376-7

The author gratefully acknowledges permission to reprint the following
previously copyrighted material:
Selections from *Evolution in Action* by Julian Huxley, Harper & Brothers,
1953. Reprinted by permission of A. D. Peters & Co. Ltd., London.
Selection from *The Heavenly City of the Eighteenth Century Philosophers*
by Carl Becker, 1932. Reprinted by permission of Yale University Press.
Excerpts from *The Great Chain of Life* by Joseph Wood Krutch, 1957. Re-
printed by permission of Houghton Mifflin Company.
Selections from *The Torch of Life* by René Dubos. Copyright © 1962 by
Pocket Books, Inc. Reprinted by permission of Simon & Schuster.
Selections from *Science and the Modern World* by Alfred North White-
head, 1925. Reprinted by permission of Macmillan Publishing Company.
Excerpt from *Mysticism and Logic* by Bertrand Russell, 1903. Reprinted by
permission of George Allen & Unwin (Publishers), Ltd., London.
Selections from *The Phenomenon of Man* by Pierre Teilhard de Chardin.
English translation by Bernard Wall, 1959. Reprinted by permission of Har-
per & Brothers (United States rights) and William Collins & Co., Ltd., London
(British rights).
Excerpts from *Problems of Life: An Evaluation of Modern Biological and
Scientific Thought* by Ludwig von Bertalanffy, 1952. Reprinted by permission
of Pitman Publishing, Ltd., London.

*This book is dedicated to my husband because
(in the words of West Side Story) we have
"made of our lives one life, day after day one
heart, one mind." The thoughts, enthusiasms,
and strengths that have flowed from this union
are different from those which might have
been mine alone—just as the purple sea fan
swaying in sunlit waters differs from the single-
celled algae that inhabit its lacy form or from
the solitary polyp growing in semi-darkness on
the ocean floor. Through symbiosis a new
whole has been created with characteristics
transcending those of its individual parts. It is
impossible to tell where the contribution of
each partner ends and the qualities of the
whole begin.*

CONTENTS

7

INTRODUCTION

*The creation . . . is not an event which hap-
pened in the remote past but is rather a living
reality of the present. Creation is a process of evolution of which man is not merely a witness but a
participant and a partner as well.*

THEODOSIUS DOBZHANSKY
The Biological Basis of Human Freedom

Into this mysterious universe we are all plunged at birth with
no set of instructions, no maps or guideposts. The questions of
meaning, purpose, and direction remain with us all of our days.
Although they may be hidden beneath the blanket of everyday
concerns, they emerge now and then, unbidden—when we look
into the grave eyes of a newborn child or stand beneath the
canopy of stars in the stillness of a summer night.

In every field and woods, in every season of the year we see
intricately designed artistic creations—a rosebud opening, a butterfly emerging from a chrysalis, a snowflake drifting on the
winter wind. Countless examples speak of an underlying force
responsible for designing a world of order and beauty. Yet we
cannot close our eyes to the ugliness, the pain, and cruelty that
are present, too. The cuckoo hatchling pushes the baby robins
out of their nest, the larvae of the ichneumon fly slowly eat out
the intestines of the living caterpillar, and human beings in
torture chambers ram pressure hoses down the throat and red-
hot pokers up the anus of their victims. How can we reconcile

9

the presence of such evil with the belief in a purposeful and ordered universe?

Attempts to resolve this paradox have taken many forms. Religious faiths have postulated that the universe is controlled by suprabeings. These deities—both good and evil—are in constant conflict. The Greeks had a whole pantheon of warring gods and goddesses. The kindly works of Zeus could be counteracted at any moment by Pluto, lord of the underworld, and Hecate, powerful in heaven and earth. Norsemen worshipped Thor, the thunder god who was both benevolent and threatening, bringing storm and destruction.

Christians believe in the presence of an evil force personified in the Devil, a force almost as powerful as God himself. The Devil brings with him all manner of temptations and has caused man's fall from grace. In the rest of nature, evil is an illusion reflecting man's incomplete understanding. What appears to be unjust or cruel, when viewed in a wider context may be seen to work for the greatest good in the end. The ways of God are beneficent even though they are inscrutable.

But as knowledge of the universe has grown, some people have found it more and more difficult to accept these explanations; many have abandoned traditional doctrines and have looked for answers in the understanding achieved by science.

The principles of classical dynamics discovered by Galileo and Newton postulated that immutable laws control the motion and interaction of material objects, large and small, everywhere in the universe. These concepts led to a mechanistic interpretation of nature in which everything that occurs today is seen as the inevitable consequence of actions that occurred in the past, and the future is the inevitable consequence of the present. Every individual, every atom, or star is just another cog in the great machine; therefore, free will and morality are figments of man's imagination. If such a universe were divinely created, it must be the result of an act that occurred at the beginning of time, and the end must have been foreseen in the beginning. God set the wheels spinning and then retired, allowing all the consequences of the act to materialize in the fulness of time. But in this case how do we account for the presence of evil, which also must have been foreseen and could have been prevented by a just and all-powerful Creator? George Bernard Shaw spoke of "the theology of the women who told us that they became atheists when they sat by the cradles of their children and saw them strangled by the hand of God." The most obvious answer to the dilemma is to remove the concept of God from the scene entirely, leaving the picture of a purely mechanical universe grinding out the inevitable consequences of the laws of matter in motion. This mechanistic outlook dominated the scientific world-view of the nineteenth century.

Then toward the beginning of the twentieth century physicists became aware of the fact that some degree of indeterminism exists at the base of natural phenomena. It is impossible to know simultaneously the exact velocity and position of very small elementary particles and, therefore, the action of these particles cannot be precisely predicted. The laws of physics appear to be iron-clad only because large numbers of events are involved at the atomic and subatomic levels; the randomness averages out, and the "laws" that emerge are simply statistical in nature. This concept of a universe spawned by chance allows some latitude for free will but denies all possibility of purpose or direction. The vestiges of form and order to which we have attached so much significance are only accidental accumulations of matter—nothing more than flotsam thrown up from the restless sea of random events over vast eons of time.

Carl Becker described this twentieth century world-view in these words:

> Edit and interpret the conclusions of modern science as tenderly as we like, it is still quite impossible for us to regard man as the child of God for whom the earth was created as a temporary habitation. Rather must we regard him as little more than a chance deposit on the surface of the world, carelessly thrown up between two ice ages by the same forces that rust iron and ripen corn. . . . The ultimate cause of this cosmic process of which man is a part, whether God or electricity or a "stress in the ether," we know not. Whatever it may be, if indeed it be anything more than a necessary postulate of thought, it ap-

*Nothing but flotsam thrown up from the restless sea
of random events over vast eons of time?*

Plate 1
Whirlpool in the Bay of Fundy.

Plate 2
Whirlpool Galaxy in Canes Venatici (Lick Observatory Photograph).

pears in its effects as neither benevolent nor malevolent, as neither kind nor unkind, but merely as indifferent to us. Man is but a foundling in the cosmos, abandoned by the forces that created him. Unparented, unassisted and undirected by omniscient or benevolent authority, he must fend for himself, and with the aid of his own limited intelligence find his way about in an indifferent universe.

Such is the world pattern that determines the character and direction of modern thinking. The pattern has been a long time in the weaving. It has taken eight centuries to replace the conception of existence as divinely composed and purposeful drama by the conception of existence as a blindly running flux of disintegrating energy. But there are signs that the substitution is now fully accomplished; and if we wished to reduce eight centuries of intellectual history to an epigram, we could not do better than to borrow the words of Aristophanes, "Whirl is king, having deposed Zeus."

Although this nihilistic interpretation of nature is widely accepted in the scientific world, most nonscientists find it even less attractive philosophically than the mechanistic view which it replaced. Unable to argue directly with the scientists, they cannot deny their own intuitive perception of a creative force pervading all nature. So they have turned their backs on science, espousing mystical, supernatural, existentialist, and transcendental ways of interpreting the experience of existence.

A few scientists and philosophers of science never did believe the world-view described by Becker. James Jeans declared, "the universe shows evidence of a designing or controlling power that has something in common with our own minds." Albert Einstein said he could not believe that God plays dice with the universe. Huston Smith writing in 1961 said: "The change from the vision of reality as ordered to unordered has brought Western man to as sharp a fork in history as he has faced. Either it is possible for man to live indefinitely with his world out of focus, or it is not. I suspect that it is not, that a will-to-order and orientation is rather fundamental in the human makeup. If so, the postmodern period, like all the intellectual epochs that preceded it, will turn out to be a tradition to a still different perspective."

Now in 1985 the changeover has still not occurred. Steven Weinberg, physicist and author of a recent book on the origin of the universe, did not find any solace in the theories he presented. "The more the universe seems comprehensible," he wrote, "the more it also seems pointless. But if there is no solace in the fruits of our research, there is at least some consolation in the research itself. . . . The effort to understand the universe is one of the very few things that lifts human life a little above the level of farce, and gives it some of the grace of tragedy."

Ironically, while the main stream of scientific thought continues to restate its denial of meaning or purpose in the cosmos, more and more remarkable facts have been emerging from cosmology, geology, and the study of life—facts that seem to point in a new direction. In the past few decades our concepts of time, matter, and process have undergone important transformations. The latest ideas emerging from modern science suggest a new solution to the cosmic paradox.

As we look farther out into space, back into the dim reaches of time, and down into the smallest units of both living and nonliving matter, it is apparent that formative processes have been at work since the beginning and are very active today. The universe is not a static phenomenon; it is changing and evolving, still in the process of becoming. The evidence from cosmology indicates that the universe began as a disorganized cloud of matter and radiation expanding into space. Since that moment trillions upon trillions of stars have coalesced out of the original chaos and have become organized into great globular clusters and the vast swirling pinwheels of spiral nebulae. The transformation process is still going on around us. Stars are being born, the galaxies are evolving, and somewhere in the uncharted paths of the night sky another planet like earth may at this very moment be condensing from a clump of cosmic dust. The advent of life, its spectacular growth and evolution, provide the most striking demonstration of the potential hidden deep within the very substance of the universe.

I postulate that we are witnessing—and indeed participating in—a creative act that is taking place throughout time. As in all such endeavors, the finished product could not have been clearly foreseen in the beginning. We know from the experience of many artists, poets, and musicians that creativity is an uncertain reaching out, a growth, rather than a foreplanned or preconceived activity. It starts with something as vague as a need, an inspiration, a yearning. Poet Stephen Spender speaks of "a dim cloud of an idea which I feel must be condensed into a shower of words."

In all creative work there is a very close relationship between the act of expression and the medium that the artist has at his disposal. The idea or inspiration becomes clarified as it is expressed, and the means of expression helps to delineate the final form. Michelangelo has been quoted as saying, "I do not carve a statue out of marble; I release the form that is within." Brewster Ghiselin, describing the experience of composing a poem, speaks of "one of those difficult, time-consuming researches into the relation between the depths of the excited mind and the possibilities of the medium, whereby the worker little by little shapes his structure and clarifies his creative intention. For that intention is not ever quite clear in the beginning; it only becomes so upon the completion of the poem. The poem is its only exact and explicit definition."

The evolution of the universe which we observe going on about us and throughout the history of the universe is just this kind of activity. Involving change and growth, it proceeds by trial and error, rejecting and reformulating the materials at hand as new potentialities emerge. The unsuccessful experiments are weeded out by natural selection and better integrated forms take their places. Little by little the process is advanced and the work of art begins to take shape. Like Stephen Spender's dim cloud of an idea, the disorganized plasma of the early universe has been condensing into a shower of things: white dwarfs and red giants and black holes, tulip trees and bluebirds and daffodils, snowflakes and sapphires, the intelligent dolphin and the inquiring mind of man. But as long as the creative process is incomplete there are inevitably many imperfections present at any one time. It is understandable that evil and failure are still ubiquitous in the world around us.

Many interesting insights have contributed to this cosmic view: a new perception of the relationship of time and universal change, a new evaluation of the role of chance in the birth of stars and the formation of the building blocks of life, a new understanding of the way complex organisms are constructed from simple ones and how communities of living things may be higher organisms in the process of becoming. Like tiny beams of light shining in a darkened room, no single insight is enough to disclose the true nature of this place in which we find ourselves. But together they have illuminated the dim outlines of its furnishings, its shape, and its boundaries. They have revealed intimations of an orderly design, suggestions of work-in-progress which had not been apparent before, and have made possible a new interpretation of this awesome, beautiful, but still unfinished universe.

TIME: THE REJECTED DIMENSION

So little do we understand time that perhaps we ought to compare the whole of time to the act of creation. . . .

JAMES JEANS
The Mysterious Universe

Long ago, in the infancy of mankind, a devastating discovery was made. Behind the endless golden cycle of days—dawn and dusk, winter and spring—that return with such reassuring rhythm, there is a slow but relentless change. Each seemingly perfect cycle does not quite bring the past; a barely perceptible aging occurs in each individual day by day, more perceptibly year by year. The prime of life is reached and passed; the strengths and beauties of youth bloom and begin to fade. As each year follows on the heels of the departing year, the ugliness and indignities of old age grow ever more prominent. No one is spared and many are moved like Yeats to cry out:

> *I spit into the face of Time*
> *That has transfigured me.*

Finally death waits and, like every generation that went before, we rebel against the cruel necessity of ultimate extinction. It is no wonder that human beings have always feared and hated time. They have sought ways to conquer it, to deny its reality or reduce its status to a fleeting phenomenon that will soon be swept into the limitless sea of eternity.

A denial of the finality of death is central to most of the world's great religions. In ancient Egypt elaborate and expensive mortuary rituals were believed to assure the departed of eternal life. Wealthy Egyptians laid up rich stores of food and treasures to provide for themselves in future life and to offer as sacrifices to the gods. They built massive tombs to protect their carefully embalmed mummies from further deterioration. Artistic skill was lavished on the interior walls of these tombs, depicting ceremonies to honor the gods and inscriptions of religious texts to guarantee the favorable reception of the departed. The pharaoh was pictured as joining the sun-god in his celestial boat on its daily passage across the sky. Thus he would participate in the eternal cycles, released forever from the menace of time.

In one form or another life beyond the grave is promised to Hindus, Buddhists, Moslems, and Christians. For Hindus and Buddhists there is reincarnation until one is set free from the wheel of life, united with Brahma or received into Nirvana. For the faithful Moslem and Christian there is immediate transmutation into life everlasting. Since no one knows for certain what lies on the other side of death, faith can satisfy this most fundamental human need—to overcome death and thereby conquer the most terrible consequence of time.

The denial of change, however, is somewhat more complicated. Simple observation confirms the fact that change does occur throughout the natural world. All living things age and grow old; iron rusts; wood decays. Even the solid earth itself undergoes transformation. Sometimes it shakes and trembles underfoot. And mountains erupt, pouring forth smoke and ash that blanket the countryside and bury the habitations of man. In all these aspects change is destructive; to primitive peoples it seemed to be intimately associated with deterioration. The positive aspects of change were not recognized because an understanding of them requires a much longer historical perspective, a more detailed knowledge of the physical world. Such insights lay far in the future.

The challenge for early cultures was to interpret the changes wrought by time in a way that made them less threatening, to reduce their importance, even—and this did take a stretch of the creative imagination—to deny the reality of their existence.

The philosophy of classical India came close to achieving a complete negation of time. It postulated the existence of an eternal and changeless reality behind the procession of everyday events, the ceaseless coming and going of the seasons. Cyclical phenomena occupied a transcendental position in this philosophy. Endless repetitions of the wheel of existence were believed to reproduce the past in all its detail. The small annual cycles that bring back the same procession of constellations across the night sky, the same recurrence of life in the spring, were encompassed in a great cosmic cycle. Four thousand million years constituted a single "Brahma day" that dawned with re-creation, ended with dissolution and reabsorption of all the phenomenal world into the absolute. Thus the changes brought by time were wiped out with each new "day."

Greek philosophers, possibly inspired by Hindu thought transmitted through Babylonian culture, developed similar cyclical models of the universe. One of the earliest thinkers, Heracleitus of Ephesus, suggested that cosmic history runs in cycles, beginning and ending in fire. Yet, ironically, Heracleitus emphasized the importance of change. Only flux and becoming are real, he said. "You cannot step twice into the same river; for fresh waters are ever flowing in upon you." Everything that appears to be static really involves continuous movement. If all movement ceased, the universe would collapse into nothing (a conclusion that would not be denied by modern cosmologists).

Other Greeks, however, adopted a much more static worldview. Plato taught that the true essence of reality consists of changeless, timeless forms. It is only their shadows or reflections—pale, fleeting images—that exhibit movement and assume different aspects with the passage of time. Plato's universe was created by a God whose existence transcended time. He was eternal, not in the limited sense of having no beginning or end, but in the sense of existing outside the realm where time has any meaning. The physical world created by God was a series of crystal spheres nested inside each other; each approached in varying degrees the perfection and timelessness of the Creator. This model of the universe was taken over and further developed by Plato's successors: Aristotle, the Stoics, and the Neoplatonists.

The Greeks had observed that stars move across the sky, describing circles every twenty-four hours. The planets, moon, and sun also move in predictable but more complex paths. The cosmic model must account for these empirical facts; it cannot be completely static and changeless. Motion must be acknowledged as an intrinsic attribute of the world system; however, the uniform rotation of perfect spheres produces no absolute change when the motion is averaged out over time. In Aristotle's universe the outermost sphere carrying the "fixed stars" was in a state of perpetual uniform rotation. It was composed of incorruptible matter and was the nearest thing to perfection in the created universe. Inner spheres carried the planets, the sun, and the moon, which were also everlasting and incorruptible but the motion of their spheres was more complicated and, therefore, less perfect. The smallest sphere of all, at the very center of the universe and farthest from God, carried the earth. In this sphere all substance was subject to generation, change, and decay.

Events on earth were believed to be influenced by the positions of the other heavenly bodies. Just as the sun causes our day and night, our winter and summer, the planets, too, were believed to exert their influences. Astronomers calculated that approximately every 36,000 years the planets return to the same configuration, and this period of time was known as the Great Year—a recreation of the "Brahma day" on a different temporal scale. The Great Year had seasons like a normal year. It had a Great Summer when the earth became so hot that devasting fires raged, and it had a Great Winter when catastrophic floods swept across the land.

This theory of the Great Year was elaborated by the Stoics and Neoplatonists. The most extreme version was developed by the Stoics, who held that all activity on earth, including the events of human history, was precisely repeated in each cycle. Socrates had drunk the hemlock cup at the same moment in every Great Year and would do so throughout endlessly recurring cycles in the future. In fact every event, no matter how small, is infinitely repeated. By this sophistry the changes wrought by time were reduced to insignificant fluctuations.

Aristotle adopted a less rigid interpretation of time. He did not believe in the exact duplication of historical events. His greatest interest was in biology, and in this realm, he conceded, individual organisms must decay and cease to be. However, the species endure. All species have existed forever and will exist for an infinite time in the future. None has ever come into existence or been totally destroyed.

These static Greek models of the cosmos deeply influenced Western thought, and echoes of them are still evident in the cosmology of the present day. Christianity, however, furnished a new and quite original solution to the problem of time. The reality of change was acknowledged, but it was limited to a relatively brief period, a small slice of time carved out of eternity. Human history had a beginning with Adam and Eve, a moment coinciding closely with the birth of the earth itself. After a brief testing period the whole world of flux, strife, and temptation will be brought to an end with the Second Coming and the Last Judgment, when mankind will be received into eternity and time itself will become meaningless. Between these two historic points a linear progression of events leads to the realization of God's purpose, the deification of man.

Thus time for the Christian is invested with profound significance. It represents the unfolding of a great plan in which his own individual destiny has a minute but essential place. The sequence of events is always unique, irreversible, and decisive. The future is open because it is affected by the actions of each individual, who can advance or retard the realization of the goal. To borrow the words of Joseph Needham, the world process is "a divine drama enacted on a single stage with no repeat performances."

Medieval man was constantly reminded of the passage of time by the tolling of church bells, recording the hours, calling the faithful to Angelus, or marking the passing of another mortal being. But for the Christian, time and change were made bearable by the promise of immortality and the ever-present possibility of imminent release from the tyranny of time. The Second Coming might occur at any moment, perhaps "at even, or at midnight, or at cock-crowing." The shortness of the temporal process made the concept of historical time acceptable, cushioned as it was by eternity ever after.

Medieval philosophers were confronted with the difficult task of reconciling these two fundamentally different world-views: the static Greek model of endlessly recurring cycles and the Christian belief in a linear progression of events which could not be reversed or turned back on itself. The authority of Aristotle was deeply respected but, on the other hand, the teachings of the Church were held sacred.

The compromise was achieved by dividing reality into two arenas. The earth, lying motionless at the center of the universe, was composed of humble matter, subject to deterioration and decay. It was the site of strife and conflict, the place where the events of history occurred. But the earth was surrounded by an Aristotelian universe consisting of nested spheres that bore the stars, the sun, the planets, and the moon on their backs. These spheres were composed of an ethereal substance that never experienced alteration. Aristotle's view that all species have existed since the beginning of time and have not undergone any change was modified to accord with the biblical story that on the fifth day God created every living thing that moved in the water, on the earth, and in the air. Thereafter all species had brought forth according to their own kind. Thus, even in the little segment of space-time where real change was acknowledged, many aspects were believed to be immutable.

As the centuries passed, more detailed knowledge about the positions of the stars and planets accumulated and the Aristotelian model of the universe was gradually modified to accommodate these facts. Spheres were added within spheres; the simple rotational motion became a complex pattern of cycles and epicycles. By the beginning of the seventeenth century the theory had become so complicated that it required 80 spheres, and still it did not completely explain the observations. In spite of all its shortcomings, however, belief in this theory persisted. It was held even long after a simpler and more elegant theory was proposed by Copernicus in 1543. The older model was supported by tradition, by the continuing prestige of Aristotle, and—most of all—by the fear of destroying the neatly compartmentalized world-view that reduced time, change, and movement to a minor role.

The Copernican theory, on the other hand, set the earth adrift; it was no longer the motionless center of the universe. The invention of the telescope revealed even more disturbing facts. The planets and the moon were found to undergo changes, as evidenced by the phases of Venus and the moons of Jupiter, and our own moon was discovered to have mountains and valleys like the corruptible substance of the earth.

More evidence of important long-term alterations was discovered by naturalists and geologists studying the crust of the earth. Fossils that resembled fish or seashells were found on mountaintops, and distinctive marine-type layers were uncovered on the continents. These were unsettling revelations indeed. The history of the attempts to fit the new geologic facts into an acceptable conceptual framework is an interesting demonstration of the way scientific ideas are molded by social, religious, and philosophical beliefs. Contrary to popular opinion, theories are not always formed by a simple, dispassionate evaluation of the facts. The same set of observations can be interpreted in many different ways. The controversies concerning the nature of fossils and the history of the earth illustrate this point.

Fossils have been the subject of interest and discussion since very early times. These distinctive and often beautiful stones are mysterious in many ways. Although a great many of them bear obvious resemblances to parts of living things—shells, ferns, tree trunks, bones of animals, even sharks' teeth—they are composed of stone instead of organic substance. Aristotle and the Neoplatonists attributed the presence of these "figured stones" to the action of a molding force that governed the growth of living organisms but also operated within the earth itself. This

Greek explanation was carried forward into European philosophy. As late as the seventeenth and eighteenth centuries naturalists were expounding the view that fossils were not remains of living things at all. They had grown within the earth from a characteristic "seed" that contained the potential of a specific form. If the seed of a clam, for example, were deposited in the crevice of a rock, a clam-like fossil might grow from stony material and would resemble in shape the clam that grew from living tissue. These explanations and others like them held sway for centuries until the evidence disproving them became overwhelmingly great.

Once the idea was accepted that fossils were actually the remains of ancient organisms it was evident that very odd creatures had inhabited the earth in former times, creatures quite different from those which we see on earth today. These discoveries, taken at face value, might have suggested that species die out and new ones make their appearance, but this conclusion contradicted the belief that all existing species came into being on the fifth day of creation and had not changed their form since that time. Robert Hooke, a bold thinker among early geologists, did make the revolutionary suggestion of extinction of species, but other scientists looked for an explanation that would be in accord with accepted doctrine. Perhaps, they suggested, creatures like these really do exist somewhere on earth in places not yet explored.

The location of fossils was another enigma. How could seashells have been deposited on mountaintops? The learned men had an acceptable explanation for that fact, too. The seashells and layers of marine-type earth had been carried by the waters of the biblical deluge to locations high on mountaintops and left there when the waters receded.

As more evidence accumulated, however, it was hard to deny the fact that many striking changes had occurred, not just a single flood. Whole mountain ranges had been elevated; repeated variations in sea level had left their distinctive marks upon the land. Furthermore, the recognition was gradually dawning that the earth must be much older than the age of 6000 years which had been calculated from biblical chronology. About the middle of the eighteenth century the French naturalist Georges

Buffon suggested a time scale of tens of thousands of years and even expressed the suspicion that millions of years might be necessary to allow for the deposit of the deep layers of sediment found in the earth's crust.

It was a Scot, James Hutton, who brought this matter most forcefully to public attention in his *Theory of the Earth,* published in 1788. From first-hand observation of rocks and erosion processes, Hutton concluded that long periods of time were necessary to allow for the slow accumulation of sediments and the gradual wearing down of mountains by erosion. All geologic changes, he declared, could be accounted for by the cumulative action of natural causes like those we see in action today. But enormous stretches of time were necessary to account for the landscape as it now appears. "What more can we require?" he asked. "Nothing but time."

These new insights were interpreted in diverse ways by the philosophers and scientists of the seventeenth and eighteenth centuries. Several of them adopted a dynamic view of the planet's past, but the change that they postulated involved a degradation from an originally perfect condition. René Descartes and Georges Buffon in France and Robert Hooke in England believed that the earth had originally been a hot, fluid ball. As it solidified a smooth surface formed; further cooling and collapse of the crust produced a deteriorated, irregular surface. The conviction that the earth in its present state is a broken and ruined ball of matter was so widespread in Europe that prominent landscape features were looked upon with revulsion. As late as the middle of the eighteenth century Casanova reported that when he traveled through the Alps he drew down the blinds in his coach to spare himself the view of those vile excrescences of nature, the deformed mountains.

Although Hutton was one of the first scientists to sense the vastness of geologic time, he espoused a cyclical view of earth history. A state of dynamic equilibrium is maintained, Hutton declared, between heat from inside the planet and external cooling processes. New layers of rock are constantly being formed out of material boiling up from the hot interior of the earth and then eroded away again. In fact, the great lengths of time involved in all geologic changes seemed to merge in Hutton's mind with the concept of eternity: "We find no vestige of a beginning," he said, "no prospect of an end."

Taking its cast of thought from pioneers like Hutton, geology did not prove to be the science that revolutionized modern man's perception of time. It was biology, the study of life, which brought forth the progressive view of earth history. Important leaders in this field began suggesting as early as the middle of the eighteenth century that life forms had evolved throughout time. Georges Buffon in his voluminous *Histoire naturelle* presented many facts supporting the idea of continuous change and extinction of earlier forms. Although Buffon failed to offer an explicit and carefully reasoned theory, his popularity and international reputation served to disseminate widely the seeds of this thought. A few decades later the theory found more concise expression in the works of Erasmus Darwin (grandfather of Charles) and Jean Baptiste Lamarck. Lamarck described life forms as constantly emerging, passing from lower to higher orders of being. His explanation of how this happened (by inheritance of acquired characteristics) was subsequently discredited, but the idea of evolutionary change was established and led eventually to the Darwinian theory.

It is an interesting fact, however, that Lamarck adopted a cyclical theory for the earth itself. Like Hutton, Lamarck believed that the planet had changed in a rhythmic manner due to erosion and deposition processes. Lamarck even suggested an early form of continental drift. The land masses may have moved slowly around the globe, he said, as erosion cut into their western shores and sediments deposited on their eastern shores built more land. Thus the continents might have encircled the planet several times, returning regularly to their original positions. Lamarck appears to have been influenced in his choice of a cyclical view of the planet's history by his religious beliefs. The concept of real, long-term changes in the earth seemed to imply imperfection in the original creation. Only a steady state could demonstrate the wisdom of God. The logic of this point of view is not perfectly clear. Why should perfection of original creation be true of the planet and not of life itself?

• • •

The great English geologist Charles Lyell carried the steady-state model of earth history to its ultimate form. Although he emphasized the need for extending enormously the length of geologic time, he maintained that all transformation had been perfectly cyclical in nature so that there had been no net change in the earth or even in the life that inhabited it throughout the whole period accessible to scientific investigation. Processes which acted in the past were the same as those that act in the present. Furthermore, Lyell postulated, even the rate of these processes has remained constant.

Lyell denied the evidence for any directional change in the nature of life as revealed in the fossil record. The change was only an apparent one, he said, because the record was incomplete. There was no proof that mammals, for example, had not existed at very early periods in the earth's history, just because fossils of their bones had not been discovered. Although extinction of species has occurred in a piecemeal fashion throughout time it is balanced by piecemeal production of similar species; so throughout time there has been a balance of opposing forces. This view was attractive to Lyell, as historian Martin Rudwick points out, "because a world in perpetual and harmonious balance could demonstrate the wisdom of the creation more effectively . . . than a world in which a temporal beginning and end could be envisaged."

The philosophical difficulty which these scientists faced was the problem of reconciling a belief in a complete and perfect creative act, which took place at the very beginning of the world, with the evidence of progressive change. But it is ironical that in defending the literal interpretation of Genesis they denied one of the most important aspects of Christian thought—the perception of time as a meaningful flow of events leading to the fulfillment of a purpose. Their theories harked back to ancient Greek ideas of endlessly repeating revolutions and a denial of the significance of time.

In the other physical sciences as well—mechanics, chemistry, astronomy—new discoveries were interpreted in ways that rejected the positive aspect of time. The discovery of the laws of motion led, as we have seen, to the mechanistic philosophy of nature in which the whole sequence of natural events was seen as the inevitable consequence of actions that occurred at the beginning of the world, and time was rendered impotent. As the distinguished French philosopher Henri Bergson said:

> In such a doctrine, time is still spoken of: one pronounces the word, but one does not think of the thing. For time is here deprived of efficacy, and if it *does* nothing, it *is* nothing. Radical mechanism implies a metaphysic in which the totality of the real is postulated complete in eternity, and in which the apparent durations of things express merely the infirmity of the mind that cannot know everything at once. But duration is something very different from this for our consciousness, that is to say, for that which is most indisputable in our experience. We perceive duration as a stream against which we cannot go. It is the foundation of our being, and, as we feel, the very substance of the world in which we live.

When the mechanistic philosophy was replaced in the early twentieth century by the view of a universe spawned by chance, the concept of time as a stream against which we cannot go was reinstated. It was seen, however, as a stream flowing always downhill, a blindly running flux of disintegrating energy. According to this interpretation the universe will expand indefinitely into a space that grows ever colder and emptier. As the density of matter decreases, it will not be concentrated enough to light the fires of the stars. The existing assemblies of matter will slowly decay following the path of increasing disorder and the universe will end in a state of nothingness.

Some contemporary cosmologists, unable to accept this prediction, are looking for ways to justify a cyclical model for the universe. They are searching for evidences of more matter hidden in the dark reaches of space. If enough matter exists, the expansion rate will gradually slow down, and gravity will begin to draw the stars and galaxies closer together again. Collapse will become more and more rapid until all matter is infinitely concentrated. Then another sudden expansion may take place. Expansion and collapse can be postulated to occur in alternating cycles without beginning or end.

But the concept of eternal cosmic cycles, harmonious and artistically satisfying as it may be, makes the whole temporal process entirely pointless. The dilemma faced by modern cosmologists is described by astronomer James Trefil: "The great debate over whether the universe is open or closed comes down to the question of whether everything will fall back on itself only to repeat the cycle, or whether the last bits of matter and radiation will disappear into a darkness that expands forever. This is, in a sense, the last, the ultimate, question of science. The cosmic switch has already been thrown; the answer, though unknown, is already ordained, and man cannot influence the outcome."

The reader will recognize several familiar themes repeated here with modern variations. Change, if acknowledged at all, must work always in the direction of dissolution. And even this negative view of the temporal process is wiped out by the assumption that the entire history of the cosmos was determined in the first few seconds, a thought that found expression a century ago in the mechanistic philosophy and was stated very eloquently nearly a millennium ago by Omar Khayyam:

> *With Earth's first clay they did the last man knead,*
> *And then of the last harvest sow'd the seed;*
> *Yea, the first morning of creation wrote*
> *What the last dawn of reckoning shall read.*

Thus we see that the remarkable flow of new information about the universe has been poured into old modes of thought. The result is a bitter brew, unpalatable to layman and scientist alike. Intuitively mankind believes in order and progress. The dichotomy between the judgment of intuition and that of modern cosmology has produced a serious conflict for mankind, a conflict that touches the very core of our being.

Fortunately, a new concept of time that offers a more optimistic interpretation of the findings of modern cosmology is already present within the accepted fold of scientific thought. The rudiments of this world-view have existed for nearly two thousand

There are cycles within cycles within cycles. But the pattern that unfolds with time does not close back upon itself; it moves outward a little at each turning.

Plate 3
North Sky about the Pole Star—Eight-hour Exposure (Lick Observatory Photograph).

Plate 4
Sundial Shell (Photograph © 1985 by Andreas Feininger).

years in our cultural heritage. Christianity's doctrine of positive change occurring throughout human history introduced a dynamic concept of time, time whose passage was seen as the realization of a purpose. This model for the progressive view of time later found expression in the Darwinian theory of evolution. The concept of organic evolution requires that time be unidirectional.

Christianity and evolution are odd bedfellows, you may say. But, actually, there is a close relationship between them. Evolution grew out of Christianity—a fact that has been forgotten in the heat of the argument about the ancestry of man and the literal interpretation of the biblical story of creation. It is possible that this theory might never have been conceived in a culture that rejected the concepts of change and progress. The men who formulated the evolutionary theory had been brought up in the Christian religion and accepted without question the idea that improvement could occur with the passage of time. This belief, coupled with their observations of the natural world, resulted in a new interpretation of the history of life. A recognition of the mutability of form replaced the old belief in the fixity of species; time was vastly extended because the slow process of natural selection needed enormous stretches of time to produce the wonderfully varied world that we see around us.

Charles Darwin did not think that his theory contradicted in any fundamental way the Christian doctrine. "There is grandeur in this view of life," he said, "with its several powers, having been originally breathed by the Creator into a few forms or into one; and that . . . from so simple a beginning endless forms most beautiful and most wonderful have been and are being evolved."

According to this revolutionary theory, time in combination with chance and the laws that govern matter has resulted in greater diversity and increasing complexity of life forms. A few biologists, transcending the confines of their own discipline, have recognized the validity of this principle in the whole of nature. Evolution is seen as a single universal process encompassing both organic and inorganic transformations.

Julian Huxley was one of the most prominent advocates of this theory: "All phenomena have a historic aspect," he wrote; "from the condensation of nebulae to the development of the infant in the womb, from the formation of the earth as a planet to the making of a political decision, they are all processes in time; and they all are interrelated as partial processes within the single universal process of reality. All reality, in fact, is evolution, in the perfectly proper sense that it is a one-way process in time; unitary; continuous; irreversible; self-transforming; and generating variety and novelty during its transformations."

This dynamic interpretation of the temporal process has deep implications for philosophy and science. It is the logical culmination of the positive and unidirectional view of time which started with Judeo-Christian thought and led through the Darwinian perception of biological evolution to the broader worldview which recognizes the important role of time in the whole cosmic process.

The Jesuit paleontologist Pierre Teilhard de Chardin, inspired by this modern view of time, suggested a new synthesis of Christian and scientific thought. He believed that the entire universe from its very beginning to the present and far into the future represents a single, unfolding pattern. The most characteristic feature of the cosmic process, he said, is a tendency to "complexification." Small units tend to combine in larger structures, creating a unity in diversity which at each step opens up a new range of potentialities. Thus inorganic evolution has led to more complex organic molecules and, finally, to the first living organisms. These increasing stages of organization and order do not involve any direct intervention by God. They are the result of a tendency to complexification within matter, whose nature was created for the purpose of producing the higher levels of form leading up to man and finally to a "hyperpersonal" unity of all men.

In these philosophies the positive aspects of time have finally been recognized, illuminated by the long historical perspective science has provided and by the understanding of the natural processes that have gradually built up the various and beautiful universe. This is a truly modern world-view. As Teilhard de Chardin said: "What makes and classifies a modern man (and a whole host of our contemporaries is not yet 'modern' in this sense) is having become capable of seeing in terms not of space and time alone, but also of duration . . . and above all having become incapable of seeing anything otherwise—anything—not even himself."

This modern perception of time was set in concise physical terms by Albert Einstein. All things exist, he said, in a four-dimensional space-time continuum and these dimensions are closely interrelated; so no object or phenomenon can be properly understood by considering its extension in space alone. All things have a shape in time as well as in space. Although the Special Theory of Relativity is primarily concerned with the measurement of motion and does not directly concern us here, the recognition of the significance of time as a dimension is central to the view set forth in this book. Time is perceived as a way of measuring the progressive change that is building the universe even while we live and take part in it.

When we look back over the history of the universe with this fresh perspective we see that we must "compare the whole of time to the act of creation."

CHAPTER 2

F ROM NOTHINGNESS TO STARS AND CRYSTALS

We have assigned the wrong words for the very beginning. . . . It could not have been a bang of any sort. . . . It was something else, occurring in the most absolute silence we can imagine. It was the Great Light.

LEWIS THOMAS
Late Night Thoughts on Listening to Mahler's Ninth Symphony

Science has not yet been able to probe back to the very first fraction of an instant, the real beginning of time before the raw material of the universe, intensely hot and almost infinitely concentrated, began to expand with an impetus so great that it is still continuing to move outward. No one knows exactly what happened at time zero, or in what state the world stuff existed before the expansion began. But as cosmologists work closer and closer to the boundary they expect to find ultimate simplicity and oneness. "The present model," said theorist Alan Guth, "attempts to build the universe from almost nothing."

The event that took place at the beginning is commonly referred to as the Big Bang and is visualized as the explosion of an enormously compacted egg of matter. I prefer, however, to think of this event as the Cosmic Birth—a true beginning and not an act of destruction as implied in the metaphor of the Big Bang. Victor Weisskopf suggested that this sudden outpouring of matter and energy into space might be compared with the dramatic opening bars of a great symphony—an overflowing of power—*fortissimo*. In what better way could the world begin?

Scientists have projected the expansion back in time to one trillionth of a trillionth of a trillionth of a second after the beginning. At that time, they believe, the most elementary units of matter materialized from the undifferentiated flux of cosmic energy like snowflakes crystallizing from a cloud. These almost unimaginably small things assumed characteristic configurations that determined their identity, such as quarks (of which there are now thought to be eighteen different varieties) and leptons (a family of particles that includes the electron). They were tiny islands of form in a sea of formlessness.

This intensely concentrated "plasma" continued to spread out into space and to cool. One hundredth of a millionth of a second after the beginning the temperature had moderated to the point where quarks could come together and form larger units of matter that were able to survive the fierce bombardment of random pulses of energy.

In order to understand how expansion can provide favorable conditions for creation we should take a moment to consider the nature of heat and pressure. To the average person these phenomena are vague, subjective sensations: the warmth of the summer sun on one's shoulder, the weight of water pressing against one's body after a high dive. But to the physicist heat can be precisely defined as the average kinetic energy of molecules and atoms moving in a random way, and pressure is caused by their impact on any surface. To visualize this more vividly imagine that you and a friend are set down in a dense crowd of highly excited people all moving independently in a totally disorganized manner. The thicker the crowd the more frequently you both will be bumped and jostled by your neighbors. The faster the individuals are moving the more severe the impacts, the more likely you are to become separated from your friend, and the less likely you are to emerge unscathed from the experience. In a similar way, when the universe was extremely hot, particles that met each other by chance were unable to form lasting associations. It was necessary for the temperature of the early universe to cool to a certain critical level before quarks could meet and fuse, creating protons and neutrons—new larger units of matter.

This act of fusion was a remarkable occurrence, the first of a long, long series of fusion processes that have constructed, step by step, the whole observable universe. Fusion involves more than a simple addition of several individual units. A melding takes place, a reworking of the material to produce a new and larger whole with properties that transcend those of its parts. Each proton or neutron (known collectively as nucleons) has a complex inner structure whose patterns of motion and energy are tightly integrated, so there is no tendency for the unit to be disrupted by unbalanced inner forces. In the case of protons this process was so successful that they have been able to maintain their identity for fifteen billion years. (The age of the universe is still a subject of dispute among cosmologists. Some estimates are as low as ten billion, others as high as twenty billion years, but the median figure of fifteen is the one most commonly cited.)

It is generally believed that all the protons which exist naturally today—at least one in every atom in the universe—were created in the first one hundred millionth of a second after the Cosmic Birth. Recently scientists have been watching with sophisticated instruments for signs that the proton may, in fact, occasionally disintegrate spontaneously, and they have reached the conclusion that if the proton decays at all its average lifetime must be at least 100,000,000,000,000,000,000,000,000,000,000 years—much longer than even the greatest estimated age of the universe. This extraordinarily tenacious unit of matter serves as the nucleus of the hydrogen atom, the simplest and most abundant atom in the cosmos.

At three minutes past zero the temperature of the universe had abated considerably—although it was still seventy times as hot as the core of the sun. Then it was "cool" enough for simple compound nuclei to form. The union of one proton and one neutron formed deuterium; one proton and two neutrons made tritium, atomic nuclei that are not very stable. However, the fusion (through several intermediate steps) of two protons and two neutrons produced a finely balanced unit, the helium nucleus. It is an extremely long-lasting form of matter and, like the proton, it is one of the fundamental building blocks of nature.

In one way, however, these and all other atomic nuclei are incomplete. They are positively charged, and in order to create a perfect state of equilibrium electrons must be added. These tiny negatively charged units of matter retain a certain independence within the neutral atom while at the same time contributing to the organization of the whole. The electrons required to balance the charge on the nucleus move in precisely ordered patterns around the center like a tiny planetary system around a miniature sun. (The planetary analogy is helpful but not perfectly accurate. Electrons in orbit do not move as small discrete bodies but more like little clouds occupying the ring of the orbit.) As the electrons drop into the predetermined orbits, any excess energy which might disturb the stability of the whole is rejected, and henceforth a considerable amount of energy is needed to remove the electrons from the atom. In a very primitive sense the atom *acts* in a way that establishes its identity and preserves itself in space-time.

The neutral atom, however, occupies a much larger volume of space than the nucleus alone. If a typical nucleus were scaled up to the size of a baseball, the diameter of the atom would be more than two city blocks. This configuration, with its prodigal use of space, makes the atom considerably more vulnerable than the compactly designed nucleus to disruption by impact with random matter and energy. When the universe was still extremely hot any electrons that went into orbit around a nucleus were knocked away again by collisions; so it was not until the cosmos was almost 500,000 years old that neutral atoms could form and retain their identity. Then vast clouds of hydrogen with a small component of helium began to form throughout the infant universe.

As we look back at the first half million years of the cosmos we see that many steps in the building of form have already taken place. Each step occurred at a certain critical temperature and pressure. As soon as things *could* come together and retain their new identities for a significant length of time this condensation occurred spontaneously, in much the same way as ice crystallizes in water whose temperature has dropped below freezing.

But how can we explain the fact that matter arrays itself in these successive complexities of form—that quarks and leptons suddenly materialize out of the cosmic flux—and these fuse into nucleons—and nucleons combine to create compound nuclei? Logically we would expect that the cosmic expansion would have distributed the hot, undifferentiated plasma of matter and energy ever more evenly and more diffusely throughout space. Although random motion would have brought together many different configurations of cosmic material, these associations would have been quickly dispersed by the great movement outward into space.

The facts all fall into place, however, if we make the assumption that there is a natural tendency for self-organized wholes to form. The wholes retain their identities, return to maximum stability after they have been disturbed, and even to a certain degree regenerate their forms when these have been fractured. These qualities are apparent in living matter, and the word *organism* has been used to define the self-regulating units that possess these characteristics. But units of inorganic matter—the proton, the helium nucleus, the neutral hydrogen atom—also possess these same qualities on a more primitive level. In view of these similarities, it seems reasonable to suggest that the concept of organism should be extended to designate all those assemblies of matter (both living and inert) which exhibit the characteristics associated with wholeness. This use of the term is not new; it was proposed more than half a century ago by the great British philosopher Alfred North Whitehead. "Biology," he said, "is the study of the larger organisms; whereas physics is the study of the smaller organisms." Today geochemists and cosmologists are find-ing characteristics of self-organization in very large units of mat-ter also: the earth, the stars, the galaxies. So the phenomenon of organism must not be limited by size or level of complexity. Each self-organized unit possesses the innate tendency to preserve and extend its own existence, thus increasing the total amount of Form as measured in space and time. I use Form with a capital F to mean the sum of all the organisms present at any given mo-ment in the universe.

Traditional science, of course, has its own way of explaining the emergence throughout time of discrete units of matter. Physi-cists attribute this result to the action of three fundamental forces of nature: gravitational, electromagnetic, and "strong" forces. (A fourth one, the "weak" force, is also postulated but this one does not contribute directly to the formation of matter.) The first three are believed to exert pressures that force together and hold together units of matter: the strong force binds the protons and neutrons within the nucleus; electromagnetic force holds to-gether the nuclei and their orbiting electrons; gravity constrains larger units of matter, causing the planets to swing in elliptical paths around the sun and making stars cluster together in galax-ies like the great rotating pinwheel of the Milky Way. Why these forces exist and where they come from is at present beyond the power of science to explain. They are simply assumed to be given, just like the matter and energy which they control.

Even with these assumptions, however, cosmologists are ex-periencing considerable difficulty in imagining the exact balance of expanding and contracting forces that could have led to the universe as it is today. A cosmos that has expanded for fifteen billion years without either collapsing from gravitational con-traction or diffusing into infinity would have required a set of conditions so narrowly defined as to be almost unimaginable. The odds of achieving that kind of precise expansion without introducing additional assumptions, said one physicist, "would be the same as throwing an imaginary microscopic dart across the universe to the most distant quasar and hitting a bull's-eye one millimeter in diameter."

Another way of interpreting these same facts is to recognize that these forces are manifestations of the whole-making tendency in nature. The amount of strong force, for example, binding together the components of a nucleus, is a measure of the degree of stability and successful integration achieved by that particular organism. Gravity can be described as the attraction of particles coming together to make a larger whole. The unifying forces are not imposed from outside, but are located within the units which they help to create and which they maintain. Therefore, they can be considered characteristics of matter interacting with matter. The formation of successively larger and more elaborately structured organisms can be said to be spontaneous because it is produced naturally by internal causes rather than by external influences.

As the expansion of the universe continued, the cosmic temperature cooled, the average velocity of particles slowed down, and stars began to form. The birth of a star is one of the most beautiful examples of matter spontaneously building larger units of form. When atoms present in the clouds of particles are concentrated enough so that mutual attraction can overcome their random motions, this force takes hold. And the process feeds upon itself: as more matter is drawn in, the attraction becomes stronger and atoms from greater distances are pulled into the developing maelstrom. They fall toward the center, their motions accelerate, and they become more concentrated. When the heat and pressure reach a critical point, hydrogen atoms in the core of the developing star begin to combine and react in ways that ultimately produce atoms of helium. The amount of matter-energy needed to build this more highly structured atomic form is slightly smaller than that contained in the four hydrogen atoms that enter into the reaction. Most of the excess is rejected as radiant energy. (This is the same process that fuels the hydrogen bomb.) In a star this energy travels toward the surface, increasing the temperature of the stellar interior and finally flowing outward into the universe. The star has been turned on. Its light starts on an immense journey across the vast expanding envelope of space, carrying the message that here in the heart of this new concentration of matter the simplest atoms are being fused into larger ones. Stars are centers of creativity, gradually changing the nature and composition of the cosmos.

The making of helium nuclei is an important step in the building of even higher levels of organization. From hydrogen and helium concentrated in the hot stellar interiors, many of the heavier atoms are forged. For example, when temperatures and pressures reach a critical level three atoms of helium join together to create a carbon atom. The amount of matter-energy stored in this unit is almost twelve times the amount contained in the hydrogen atom. Almost, but not exactly. Here again a small amount is left over. Rejected in the form of energy, it increases the temperature of the stellar core and more radiance flows out into space.

As heat continues to build up, the greater average velocity of the atoms in the core helps to counteract the ever-increasing forces of attraction—the "gravity" drawing the atoms more tightly together. Once more a threshold level of temperature and pressure is reached. Four helium atoms combine through several intermediate reactions, producing the oxygen atom. By the time this happens, however, a relatively small star may have consumed most of its hydrogen. Because less heat is being generated in fusion processes, the core condenses; the star becomes unstable and finally erupts with great violence, scattering the atoms that it has fired far out into surrounding space. Here they join the clouds of almost pure hydrogen. As these "nova" explosions take place, the gases of space gradually become enriched with helium, carbon, and oxygen. Now when concentrations occur by random mixing in these clouds, they become the nuclei of new "second-generation" stars, whose initial composition includes a proportion of the more complex atoms. The presence of these atoms in the new-generation stars makes possible a whole new set of fusion reactions which, step by step, build the heavier atoms. The addition of one helium atom to carbon makes oxygen, one more helium atom makes neon, two more makes magnesium, three more silicon, four more sulfur, and so on up through iron. This whole series of elements, whose atomic weights are very nearly multiples of the helium atom, are quite stable and abundant in the universe. More complex sets of reactions produce the intermediate atomic forms. These are rarer and tend to be relatively short-lived.

Thus we see that the stars are cosmic alchemists, making in their cores all the elements up through iron from simple raw materials: neutrons and electrons, hydrogen and helium nuclei. Certain formative principles emerge as we observe the alchemists' finished products. Some types of structure are preferred over others. Like human artists designing a cathedral, a temple, a statue, or a construction in space, nature places a high value on symmetry. Even numbers of protons, neutrons, and electrons are most favorable for achieving a symmetrical atomic structure, a finer balance of forces in space. Thus symmetry contributes to the stability of the finished form.

For hundreds of millions of years a rough state of equilibrium is maintained between the forces of mutual attraction drawing the material of the star more tightly together and the outward forces caused by the release of radiation in fusion reactions. But fusion of the heavier atoms gives off less energy than the processes that created the lighter ones. Eventually an energy crisis occurs in these second and later generation stars, and contraction is unimpeded. As matter falls inward, tremendous temperatures and pressures are built up. The matter is bounced back into space in a tumultuous, glowing, shock wave as the star explodes in a dazzling display of force. Within this fiery cloud the final elements are forged—the atoms heavier than iron and the more exotic forms of the lighter atoms. Many of these are radioactive. They disintegrate gradually, dropping off fragments of matter and energy as they fall back to simpler and more firmly integrated atomic forms. Lead and bismuth, with masses slightly more than 200 times that of hydrogen, are the heaviest stable elements.

There appears to be a natural limitation to how much matter-energy can be effectively compacted into an atomic nucleus. Beyond that point the unit is less "fit," and eventually nature makes use of another principle in building more elaborate organizations of matter. Instead of increasing the size of the nucleus and the number of electrons orbiting it, whole atoms (or almost whole atoms) are joined with other atoms in an open structure. By spreading matter out more diffusely in space a vast new potential for increased complexity and variety is achieved. In the tightly compacted atomic nuclei the possibilities for different configurations are limited, but when matter is distributed over a larger volume, the parts can be arranged in many different ways.

These molecular designs are also based on the principles of number and symmetry that prevail at the more basic atomic level. There is an emphasis on orderly patterns of motion reminiscent of a dance like a formal minuet, or a Scottish eightsome reel in which four pairs of partners execute perfectly coordinated movements according to the beat of the music and the rules of the dance. Eight is a favorable number of participants, providing symmetry and balance. If there are nine dancers one will be odd man out. If there are seven, the reel cannot be performed with the same elegance and perfection.

In a similar way certain numbers and arrangements are favored by nature for the building of atomic and molecular forms. Eight is one of several preferred numbers for electrons orbiting an atomic nucleus. The atom of sodium, for example, has nine electrons. Eight of these share a single orbit. The ninth occupies an outer orbit all alone and is only lightly bound to the nucleus, even though its charge is needed to achieve electrical neutrality. The chlorine atom, on the other hand, has seven electrons, balancing the seven positive charges on its nucleus. When sodium and chlorine atoms approach each other, a higher degree of symmetry is achieved if the chlorine atom borrows an electron from the sodium atom, so that each atom has eight electrons. This exchange occurs even though it disturbs the electrical neutrality of each of the atoms. Nature appears to be more interested in arrangement than in the simple necessities of force counteracting force. Since a negatively charged particle has passed from sodium to chlorine, while the positive charges on the nuclei remain unchanged, the sodium atom is left positively charged and the chlorine atom is negatively charged. They adhere together by "electrical attraction," or, to put it another way, they arrange themselves in a form that achieves symmetry and balance in a larger organism. Thus the molecule of common table salt is formed.

The atoms of a molecule are more loosely bound together than are the parts of their respective atomic nuclei, so the molecule can be broken apart more easily than the atoms of which it is composed. But with this sacrifice in longevity comes a great increase in variety and flexibility. An astonishing diversity can be generated by combining approximately a hundred kinds of atoms in different molecular arrangements, as we can see by comparing the objects that populate our solar system: the dark, pockmarked bodies of Mercury, the moon, and Mars; the glowing body of Jupiter shining by its own light; the soft radiant orbs of Venus, Neptune, and Uranus; the small jagged rocks of the asteroids; and the flaming streamlined bodies of the comets. Here on earth almost everything we see or touch or taste is built of molecules, from the cold granite rocks of the continents to the scalding-hot lava that pours from volcanic craters, from amorphous cumulus clouds to the exquisite detail of a butterfly's wing. Form added to form and poised upon form has produced a profusion of creations.

From molecules (and in rarer cases from atoms also) nature builds three different states of matter, each one representing a higher level of organization than the one before. In the gaseous state molecules lead an independent existence, moving on erratic paths. The average motions are great enough so that the molecules that meet by chance bounce off each other and cannot form lasting associations. But as the temperature and velocity decrease, each molecule occupies less space. They move closer together and enter what we call the liquid state. In this loosely organized assembly each molecule is affected by the presence of its neighbors. Some attraction exists between them, but the spatial configurations are constantly changing as the molecules slide past each other.

As the temperature is again reduced below a certain critical level, the molecules or atoms freeze into an elaborately designed construction in space. Anyone who has watched the intricate

Each crystal is a work of art, beautiful in its orderly perfection—an embodiment of the supremely logical structure underlying all things.

Plate 5
Snow Crystal (Courtesy of United States Department of Commerce, Weather Bureau).

Plate 6
Cornflower (Photograph 1961 by Tet Borsig).

patterning of a frost flower spread across a windowpane has witnessed firsthand the mysterious creative activity by which much of our world has been formed. Each crystal is a work of art, beautiful in its orderly perfection—an embodiment of the supremely logical structure underlying all things.

With the act of crystallization nature completes five levels of form, each one built upon the one before: (1) nucleons are created by the fusion of quarks, (2) protons and neutrons unite, making atomic nuclei, (3) electrons join with nuclei to complete the neutral atoms, (4) atoms unite with other atoms to construct the molecule, and finally (5) molecule balanced upon molecule fashions the complex spatial configuration of the crystal.

Each type of crystal has a characteristic arrangement, a pattern of atoms repeated over and over again in three dimensions, like a skyscraper with every room identical. For example, quartz crystals are made up of tiny hexagons, salt crystals of little cubes. These units are added as the crystals grow until an obstacle is encountered and growth is restricted.

The size of the crystal brings us to a level of form which can be observed with the naked eye or a low-powered magnifying glass. Some simple and interesting experiments provide insights into the process. If you dissolve salt in water and allow the water to evaporate, the salt will recrystallize and collect in the bottom of the glass. You can see that it is composed of little cubes with flat faces. Dry one of the cubes and then you can cleave it with a razor blade along planes parallel to the faces of the cube. In this way smaller cubes with flat faces can be made from the original one. If you dissolve sugar as well as salt in the water, the two will crystallize out separately. The salt crystals will contain little or no sugar, because crystals grow by the selective addition of like molecules from the solution. This selectiveness is another manifestation of nature's tendency to build ever higher orders of differentiation and form.

The experiment of growing a large and quite perfect crystal is simple in principle, although the details of the technique may require some experience. A concentrated solution of the substance to be crystallized is prepared and placed in a clean container. Then one tiny "seed" crystal of the same mineral is suspended in the liquid and the presence of this template starts the process. Working outward from this center, molecules take their places in precisely ordered arrangements. The crystal seems to grow quite gradually from day to day and yet the rapidity of the action is almost beyond human comprehension. The building of a typical crystal involves the proper placement of something like a trillion molecules an hour.

Swift as it is, the crystallization process is sometimes too slow to keep pace with very rapidly falling temperatures. For example, when hot magma erupts under water or in very cold air, the lava cools and solidifies so quickly that there may not be enough time for crystals to form. Then amorphous volcanic glass, such as obsidian, is produced. In fact, glass is made by melting silica (with other minerals) and quenching the solution so quickly that crystals cannot form.

Obsidian is a beautiful substance, remarkably free of bubbles or imperfections. It is jet black with smooth, shining surfaces. Obsidian was highly prized by primitive peoples because it could be chipped to make razor-sharp tools. The Aztec priests used obsidian knives to cut out the hearts of their living victims. And according to some authorities, before the discovery of metal in China, sharply pointed pieces of obsidian may have been used as needles in the practice of acupuncture.

This relatively formless solid state, however, is not very long-lasting. It gradually gives way to the elegant architecture of crystal structure. Very slowly, a transformation takes place, altering the arrangement of the tightly packed molecules. Small white crystals begin to form in tiny centers scattered more or less uniformly throughout the rock, creating "snowflake" obsidian. Unaltered obsidian becomes progressively rarer in rock formations as one goes back in geologic time. It is never found in deposits older than a few million years. Other forms of volcanic glass undergo a similar metamorphosis somewhat more slowly, taking perhaps thirty million years to complete the process.

Crystallization has been one of the most important processes that have shaped the world we know. The lithosphere of the earth and all the inner planets is an intricately interwoven fabric of many crystals. Far out in the solar system, ice crystals circle in rings of iridescent light around Saturn and Neptune. And much much farther out, in the dark spaces between the stars, tiny crystalline grains of star dust may hold the key to some of the most creative and mysterious processes occurring in the cosmos.

When atoms fired in the stellar furnaces are flung into space they enter a cooler environment where conditions are suitable for the formation of crystals. Minute grains crystallized from star fragments—countless trillions of them—float in what appears to be empty space, but in some places they are concentrated enough to deflect the light, creating opaque clouds of matter. Many of these crystals are sturdy enough to withstand the random impact of energetic radiation in outer space. The elegant symmetry of the crystal contributes to its stability. Large and elaborate molecules are more delicate, more vulnerable to disruption from chance encounters. The organic molecules on which life is based are extremely fragile. These forms are able to survive here on earth because our planet provides an exceptionally benign and gentle environment, sheltered by the soft layer of atmosphere from the harsh impact of high-speed particles and pulses of energy that rocket ceaselessly through the space between the stars.

For many years scientists assumed that the environment of interstellar space was too harsh to allow any but the most simple and stable molecules to survive there. However, in the last few decades, studies made with the radiotelescope have revealed the presence of many molecular forms floating in the clouds of interstellar space: water, ammonia, compounds of silicon and sulfur, and a large number of carbon compounds. It is a strange and interesting fact that interstellar molecules seem to be predominately organic. At the present writing some 57 species have been identified, and three-fourths of them contain carbon.

When these molecules are disturbed by impact they rearrange their internal organization to restore maximum stability and reject the excess energy as radiation. The wave pattern of the radiation produced by each type of molecule is as characteristic as a fingerprint. Thus astronomers can identify and estimate the abundance of these molecules in interstellar clouds.

When these calculations were first made they revealed surprising information: the molecules are present in much larger quantities than could be explained by conventional theory. Scientists had estimated that most of these molecules would be destroyed in a dozen years or so, and that incredibly long periods of time would be required for new molecules to form in space and take the place of those that had been lost. Interstellar clouds are very diffuse, with a density equal to the best vacuum ever produced in a laboratory on earth. And they are extremely cold, with temperatures ranging down to near absolute zero, the theoretical limit at which all molecular motion ceases. Under these conditions the average time required for two hydrogen atoms to encounter each other and come together in such a way that a molecular bond could be achieved between them would be more than the estimated age of the universe.

And yet the molecules are there, tumbling end-over-end in the near vacuum of space and sending out their distinctive messages. Every year several more kinds are discovered and added to the growing list. These facts cannot be denied, and ingenious new theories have been proposed to explain them. Perhaps the molecules survive longer than expected because they are shielded by crystalline grains of star dust from high frequency radiation. Perhaps these grains also act as "catalysts," speeding the formation of molecules. A catalyst acts as a guide or template which accelerates one specific chemical reaction without itself being used up in the process. For example, an oxygen atom may be attracted to the surface of a dust grain and held there until such time as another atom—perhaps carbon—wanders by. Then it, too, is attracted to the surface, and bonding between the two can occur, making the carbon monoxide molecule. Later, released from the surface, the molecule joins the vast, amorphous clouds that wander through space.

In fact, some astronomers believe there is a symbiotic relationship between interstellar molecules and the formation of stars. Radio and infrared observations show that these molecules are most abundant in the darkest regions of the clouds, where star dust is the densest. These are also the regions where stars are born. Although many details of star formation still remain obscure, it is generally believed that the presence of dust particles and interstellar molecules may help to start the collapse of interstellar gas clouds. If the cloud is too warm and too diffuse, it will not condense. But dust grains attracting other atoms to their surfaces act as catalysts, helping to build molecules, and these in turn contribute to the density in that particular region of space. They also cool the cloud by radiating energy when they have been slightly disturbed. Thus the two inhibiting conditions—heat and dispersion—are mitigated. Once the density and temperature reach certain critical values, mutual attraction takes hold and stars begin to form.

For reasons that are not completely clear, however, total collapse of the cloud at this point does not usually occur. In our own galaxy, for example, a delicate balance seems to be maintained between expansion forces and condensation in the molecular clouds. Although star formation is occurring, it is less rapid than would be expected from the temperature and density of the clouds. One intriguing possibility that has been suggested to account for the maintenance of this condition of relative equilibrium is that star formation in a molecular cloud may be a self-regulating process.

Interstellar matter is still plentiful in our galaxy, although waves of star-birth seem to sweep through the dark clouds that float in the spiral arms. The mysterious and beautiful Great Nebula near the Sword of Orion has long excited mankind's curiosity and wonder. Now with infrared telescopes astronomers can probe deep into the hidden recesses of the cloud from which no visible light escapes. There they have found a number of radiant sources that are probably newly formed protostars. In the depths of the Large Magellanic Cloud, the galaxy that lies closest to the Milky Way, a dense cluster of stars is forming in the black regions of the Tarantula Nebula. Some of these new stars are perhaps a hundred times as massive as our sun.

The hot young stars pass through their lifespans forging more and more complex atomic forms in their cores until they finally explode and spill their precious load of highly organized forms of matter into space. These, too, join the interstellar clouds where they act as raw material for a new generation of stars. As each of these stars condenses out of the cosmic cloud, smaller bodies, too, may take shape around it. Mutual attraction draws together the random bits of matter in the nebula and organizes them into planets, moons, and asteroids. In our own stellar system those bodies that are nearest the sun are composed predominately of the highly evolved atomic forms. Since these bodies have relatively small masses and therefore small gravitational fields, they have lost most of the lighter elements, such as hydrogen and helium, to space. The inner planets like our earth are composed principally of matter that crystallized from star dust, and from these elements nature has created something even more wonderful—life itself.

Thus we see that the stellar cycle of birth, development, and death is not idly repetitive. Each turn of the cycle brings new units of more highly developed order into the universe. As matter passes through the stars, cosmic space becomes enriched with more complex units of matter. At each stage of its development the universe becomes more differentiated, furnished with a greater number and variety of organisms than it had been before. *With the elapse of time Form has increased.*

Throughout all these natural processes we have observed an interesting principle at work. Wherever matter exists, other matter can take shape more easily. Many organisms act as templates for similar or even more complex ones, and the presence of matter can trigger activities that draw bits of matter together, starting the creation of something new.

From the time when the first quarks and leptons materialized out of the undifferented cosmic substance, to the present-day universe with its almost infinite variety of things—galaxies and quasars, pulsars and black holes, red giants and white dwarfs, stars and planets, and more than a million living species that clothe our planet with an ever-changing kaleidoscope of shape, color, and movement—the universe has opened out and blossomed into ever more intricate and elaborate forms.

This great flowering of creativity is still taking place all around us. The serene, star-studded heavens which we observe on a moonless night are really an arena of intense activity, surging with undreamed-of power. Most of this action cannot be detected by the human eye alone, but if our eyes could receive X-rays, gamma rays, radio, ultraviolet, and infrared, we would see our familiar quiet sky lighted up and spouting fountains of fire. The whole dome of the heavens would glow with a soft X-radiation whose source is still unknown. We would see galaxies hurling jets of matter into space and occasionally illuminated by the explosion of a dying star sending out shock waves that buffet and heat the interstellar gas to millions of degrees. Beams like searchlights would gleam from spinning neutron stars which rotate at incredible speeds. (The pulsar embedded in the brilliant cloud of the Crab Nebula spins thirty times a second; the one known as 4C21.53 is spinning more than 600 times a second.) About once a week we would see a startling burst of radiation coming from within our own galaxy. These mysterious flashes of gamma-radiation that release as much energy as billions of stars have not been explained.

When we look out into the universe with the new tools which perceive all these energies, we are awed and humbled by its size and power, by the seeming insignificance of the tiny spark of life on a small planet orbiting a second-rate star near the outer edge of a very ordinary galaxy. It seems incredible, as James Jeans once said, "that the universe can have been designed primarily to produce life like our own; had it been so, surely we might have expected to find a better proportion between the magnitude of the mechanism and the amount of the product."

But an understanding of the history of the cosmos suggests a different conclusion. The magnitude of the mechanism may be an indication of the *value* of the product. Not size nor power but potential is the true measure of importance—potential realized through the meticulous construction of higher and higher degrees of organization, from quarks to the finely wrought molecules of living things. Generations of stars were required to synthesize the elements essential to life. The time required to mold these miniature masterpieces of design was at least twelve billion years. As British physicist John Barrow remarked, "The universe *has* to be large for life to have evolved."

CHAPTER 3

THE SYNTHESIS OF LIFE

The secret, if one may paraphrase a savage vocabulary, lies in the egg of night.

LOREN EISELEY
The Immense Journey

Life is movement and color, sound and light. It is the whirr of pigeon wings at sunset, the lonely cry of a coyote on a distant hill, the flicker of fireflies on a summer night. From common, everyday ingredients like dust and stone, air and water, nature has spun these ephemeral forms, "these unique and never recurring patterns in the stream of time."

The form of a living thing is not anchored to the substance that embodies it. As in a whirlpool, the shape of the whole is retained while matter flows into it and out. The shape itself evolves with time and yet the identity of the organism is preserved; the whole time-dependent pattern is replicated and multiplied from generation to generation. The invention of life has set form free to dance through time and space, endlessly recreating its own image like a ballerina performing in a hall of a thousand mirrors.

The transformation from nonlife to life is so dramatic we can easily imagine that a vital spark must have been breathed into matter to create life, that an unbridgeable gap exists between the realm of living and inanimate matter. The experiments of Pasteur seemed to place this premise on a firm scientific basis—for every living organism there is a living precursor, he declared. But the findings of twentieth century science lead to the opposite conclusion. Although it may be impossible for life to be generated *de novo* from formless matter on earth today, this event probably did occur under different conditions on the primitive earth, and may have happened in many other places in the universe. As we have learned more about the nature of matter, there seems to be little doubt that life is a direct and probably inevitable consequence of the order of nature. We are beginning to suspect that the potential was present in matter back in the almost beginning when quarks first materialized out of the formless plasma; perhaps it was even there in the cosmic "egg of night."

Theories about the origin of life have passed through several phases, and are still in flux today. During the early part of this century and well into the 1960s most biologists believed that life was created by pure chance in the warm soup of the primordial seas. According to this theory, when the earth's first crust was formed, nearly four billion years ago, it was as sterile as the surface of the moon. The land masses were subject to wide variations in temperature and exposed to lethal ultraviolet radiation from the sun. But the oceans provided a warm temperate environment. In their depths, protected by the water medium from the harshest components of solar energy, elements were slowly agitated, meeting and parting in a random motion that continued for perhaps a billion years. Given this great length of time all possible combinations of the elements would almost certainly have occurred. So it was probable that the right molecules would meet by chance and form the organic molecules that were the precursors of life. These molecules accumulated, and the primitive seas became thick with them. The restless ocean currents, the tides, and vagrant breezes continued to agitate this rich brew for at least another billion years, bringing together in turn all possible combinations of these molecular building blocks until one day the first living thing came accidentally into being. This primitive form of life thrived and multiplied, feeding on the organic compounds in the sea around it.

In this scenario a very long period of time was required to turn an unlikely happening (the chance coming together of all the essential ingredients) into an inevitable or at least a probable event. But this requirement was no problem because time in large quantities seemed to be available. As late as 1960 there was no evidence that the advent of life on earth occurred until two or three billion years after the planet was born approximately 4.6 billion years ago.

In the last few decades, however, we have learned that life appeared much earlier on the earth. The speed of its evolution from inorganic matter is one of the most surprising recent discoveries. As geologists have identified and analyzed more and more ancient rocks, the advent of the first forms of life has been pushed back into an early period in the planet's history.

Very ancient rocks have been identified in western Greenland; radio dating has placed their age at 3.8 billion years. These rocks are believed to be part of the first permanent crust of the planet. Microscopic examination of the formation has revealed the presence of organic molecules characteristic of living organisms— carbon compounds similar to those produced by biologic activities. These findings suggest that life may have been present even at the time when the continents crystallized on the planet and the first oceans began to form.

Definite evidence of life has been found in rocks 3.5 billion years old. They contain fossil organisms that were already advanced beyond the most elementary phase. Thus the time required for chance alone to produce living matter has been reduced from two billion years to a relatively small segment of geologic time. Mechanisms other than the workings of pure chance must have been involved. Perhaps the building blocks were synthesized in the intense heat of volcanic action or lightning flashes during the chaotic years when the earth's crust was forming. Perhaps they were created elsewhere and were brought to earth by meteorites or shooting stars. These theories would have been ridiculed a few decades ago; now they are among the favored explanations of the origin of life.

It is a remarkable experience to watch a shower of shooting stars. I remember one brilliantly clear night in August when the earth was swinging in its annual path through the debris of a comet's tail, producing the Perseid shower. I was on a sailboat, riding at anchor in Beardrop Harbour on the coast of Georgian Bay. Far from the lights of any town I had an uninterrupted view of the night sky. At first I could see no movement at all in the great light-studded dome overhead. It seemed infinitely far away, eternal, unchanging—like the perfect sphere of fixed stars in Aristotle's universe. Then suddenly I saw a tiny path of fire curving down the sky as though a star had dropped from its place in the heavens and was moving swiftly toward the earth. One by one, other stars took flight—more and more until my mind and eyes were dazzled by the shower of fire falling from the cloudless sky. I imagined that I could reach up and catch a handful of fallen stars.

Of course, stars had not really left their places in the sky. The earth had passed through—and gathered unto itself—a diffuse cloud of crystals spilled from a comet's tail. Surprising as it may seem, one might touch a mote or two of this cosmic dust as it floats slowly earthward. Usually these visitors from outer space are so small that they burn up long before they reach the earth's surface, leaving nothing but a trail of light, ashes, and tiny crystals. More than a million billion of these stray bits of matter—meteoroids and fragments of asteroids as well as comets—enter the earth's atmosphere every day, distributing several tons of meteor dust across the planet's surface. And occasionally a piece large enough to survive the fiery trip through the atmosphere falls to earth.

On a bright September morning in 1969 an Australian farmer who lived near Shepparton, Victoria, went outdoors to play a game of tennis. As he stepped out of his house he was startled to see "an enormous ball of fire that suddenly whooshed across the sky." The apparition was accompanied by a loud crackling sound as though "a big heap of twigs was burning." This noise was punctuated by several explosions, and puffs of smoke were clearly visible even against the sundrenched sky. Several other observers watched this same display as the meteorite traveled from southeast to northwest and disintegrated above the little town of Murchison. Dozens of fragments were collected from this area over the next few weeks and were sent to museums and universities around the world. These stones are black or a deep reddishbrown, with a crusty exterior. When first discovered they had a peculiar odor which was identified as pyridine, a complex compound known as an "organic base." It is similar to compounds that appear in the genetic code.

Rocks like these are samples of matter that formed when the sun and its planets condensed out of the solar nebula. Most of them are very old; about 4.6 billion years have elapsed since their birth. Throughout this great length of time these small dark bodies have been wandering on lonely paths through the solar system. Many of them have circled in the asteroid belt between Mars and Jupiter until a combination of gravitational fields carried them out on paths that brought them near enough to earth to be attracted to its surface. Some meteorites are chips from much larger bodies. A rock that has been identified as a piece of the moon has been found on the glaciers of Antarctica, as well as several that may be fragments of Mars.

Meteorites are interesting for many reasons. They are the oldest pieces of matter that can be handled and examined in the laboratory. They have not been subjected to erosion from rain, wind, and flowing water as the earth has been and, therefore, they are thought to have changed much less since their formation than the crust of the earth itself.

There are several different kinds of meteorite. Some are metallic; others are stony, with a wide range of mineral constituents. One type of stony meteorite contains almost all the elements that make up the solar system as a whole, and in approximately the same proportions. These meteorites—known as carbonaceous chondrites—are believed to be relatively unaltered samples of the material that condensed from the solar nebula during the last stage of its cooling. The rock that fell on Murchison, Australia, is one of this group.

Analysis of carbonaceous chondrites turned up some fascinating information that electrified the scientific world in the 1950s. Carbon compounds, including some of the amino acids present in all living things, were identified in these meteorites. In November 1961 two biochemists announced that they had discovered microscopic-sized particles resembling fossil algae in relatively large quantities within two recently fallen meteorites.

This finding, if confirmed, would have been evidence of life on other bodies in the solar system. But the results of follow-up investigations were disappointing. It is extremely difficult to avoid introducing some terrestrial contaminants when these very sensitive measurements are made. A tiny speck of foreign substance could have accounted for the remarkable results that had been announced, and indeed, analysis performed by other scientists showed that the wonderful fossil algae were nothing more than minute fragments of something resembling Kleenex! In the wake of this disappointing episode chemists returned soberly to their laboratories and worked for many years to perfect their analyzing techniques. Now, after two decades of meticulous research, they have demonstrated beyond reasonable doubt that large organic molecules are present in many freshly fallen carbonaceous chondrites. These molecules include most of the essential building blocks of life.

Two kinds of organic chemicals—proteins and nucleic acids—are involved in the structure of even the simplest living thing we know on earth today. These are both long complex molecules constructed of simpler units. In the protein molecule the units are amino acids, which come in many different forms; twenty of them are used in building protein molecules. The nucleic acid molecules carry the genetic code of every living organism. They are constructed of three kinds of simpler units: one of two sugars, one phosphate molecule, and five different forms of organic base. With the exception of the phosphate, all of these building blocks are composed of four basic elements: carbon, nitrogen, oxygen, and hydrogen.

In 1984 scientists at the University of Maryland's Laboratory of Chemical Evolution announced that they had identified in the Murchison meteorite all five of the organic bases used in the genetic code. They and other researchers had also found in this same meteorite most of the amino acids which are the essential ingredients of protein molecules.

While this line of research was turning up evidence that the building blocks of life could have been created in many places throughout the solar system, another series of experiments was proving that the essential molecules could have been synthesized quickly and efficiently here on the primitive earth. The primordial atmosphere did not contain any free oxygen; it consisted of methane, carbon dioxide, hydrogen, water vapor, and ammonia. (These gases contain the key elements: nitrogen, carbon, oxygen, and hydrogen.) The atmosphere was constantly bombarded by energetic ultraviolet radiation from the sun and flashes of lightning. Given this combination of gases and energy, the creation of the building blocks might have occurred very rapidly—perhaps even in a flash of lightning. This was the thought that inspired the American chemist, Stanley Miller. In 1952 he performed a landmark experiment in which a mixture of water, ammonia, methane, and hydrogen was circulated past an electric discharge. At the end of just one week he found many organic compounds in his solution—even some of the amino acids were present. Apparently, large periods of time were not necessary for the synthesis of the basic molecules from which life may have been formed.

Experiments like Miller's have been conducted frequently in the intervening years, with slightly different combinations of gases and different sources of energy. These have produced similar results and have succeeded in generating sixteen of the twenty amino acids critical for life. In 1984 an experiment carried out at the University of Maryland produced all five of the organic bases of the genetic code, bases that had also been found in the Murchison meteorite.

From these two lines of evidence we can conclude that the first steps in the creation of life took place quickly and easily, perhaps in many locations throughout the cosmos. The process does not rely on the slow mechanism of chance combinations produced by random mixing alone. On the contrary, when the right conditions are present the molecules form spontaneously. The scientists working on these experiments have been impressed by the fact that the key molecular constituents seem to show an eagerness to associate in ways that lead to higher orders of complexity.

In making organic molecules like these building blocks of life, nature has invented a new and more versatile way of organizing matter. Spatial arrangement has taken on prime significance. The simpler inorganic molecules usually have only one possible arrangement of their constituent atoms, but in the more complex organic forms the arrangement of the atoms can be varied in many different ways, and the spatial configuration determines the characteristics of the compound. For example, there are three kinds of sugar—fructose, glucose, and galactose—which have the same number and the same kind of atoms in their molecules, but the arrangement of these atoms in space is slightly different in the three forms. These structural differences produce important variations in properties. A simple analogy is the arrangement of letters in a word. *God* and *dog* use the same letters but have quite different meanings.

Pattern and configuration in space take on more and more importance as we move to even higher levels of form. The single building blocks (known as monomers) are combined in long chains to make the very elaborate molecules of protein and nucleic acid. These "polymers" may contain as many as ten thousand building blocks—approximately a hundred thousand atoms.

It is interesting to see how this building process compares with the construction of the crystal, the most complex inorganic form that nature has created. In the crystal, you remember, each building block is identical, and the pattern is built up in a manner that produces symmetry in three dimensions. Polymers, on the other hand, are basically two-dimensional in design. They are created by linking the building blocks together in long chains. Furthermore, the blocks are not identical. There are a number of possible alternative forms of the monomers used in the polymer systems: five bases, twenty amino acids, and so on. Combining these alternative forms in different sequences, an almost infinite variety of molecules can be constructed, each with its own special characteristics. Protein molecules, for example, come in many different forms, each one designed to perform one of the many functions of a living organism, just as the twenty-six letters of the alphabet can be arranged in endless ways to serve purposes as diverse as a legal document or a sonnet. The nucleic acid alphabet consists of only seven letters and yet it is capable of generating a seemingly endless variety of genetic codes. In the human species the code of each individual is unique, unlike any other that has appeared on earth.

Biochemists attempting to reconstruct the conditions that could have led to the synthesis of the first polymer chains have encountered a theoretical difficulty. In order to forge such a chain a molecule of water must be removed at each link, and this reaction does not occur spontaneously in a water environment like the warm soup of the primordial seas. One possible answer to the puzzle was suggested by experiments performed in the 1960s. It was found that common clay can act as a mold or template aiding the formation of polymer chains. The molecular structure of clay consists of stacked sheets of silicates which are negatively charged and which provide an enormous amount of surface area. Certain molecules that form from amino acids and phosphates are selectively attracted to this surface and held there. Under dry atmospheric conditions the water molecules are evaporated. Then, when another molecule of suitable form happens by, it is attracted to the surface and linked with the first one, and so on. In this way long chains consisting of many molecular building blocks can be joined together on the clay surface. During the early period of the earth's history these polymers might have formed on the drying banks of a shallow bay. Later washed out by rising tides or rain, these molecules would have accumulated in the ancient seas.

If this promising theory proves to be true, the chances that life has been created in many places throughout the universe will be enhanced. Clay is derived from the combination of feldspar and water, substances that are ubiquitous throughout the earth's crust and even in the carbonaceous chondrites. They may be present on planets in other star systems and other galaxies.

Once again we see that the presence of matter encourages the formation of more complex units. Clay is matter in its most humble form—the mud of the fields that we scrape from our boots in the spring, the dirt that clings to plowshares and cakes the gardener's trowel. Yet even this most unpromising of all nature's creations may have played a pivotal role in the evolution of Form, acting as the matrix of the precursors of life.

This same principle can be observed at work in the next stage of complexification. The presence of a single strand of nucleic acid causes the creation of a complementary strand. Each link in the polymer chain selects from the medium in which it is immersed a partner whose configuration best complements its own. These two molecules fit together like pieces of a jigsaw puzzle, and the pairs are linked one after another in the chain. The completed double strand, which may contain as many as a thousand pairs, then turns upon itself, creating a long cylindrical helix and achieving a degree of symmetry in the third dimension. If we could look down the axis of a DNA molecule we would see an exquisitely wrought pattern like a piece of lace (see Plate 11).

In the process of doubling its form the molecule has grown and increased its occupancy of space. Most important of all, a new potential has been realized—the ability to replicate itself. Under certain conditions the double strand separates and the two single strands move apart. If each one is surrounded by a solution which contains a plentiful supply of the essential building blocks, a new complementary strand forms just like the original partner. In this way two identical double strands are created where before there had been only one. The organism has propagated itself, extending its orderly arrangement in both space and time.

Copies of these delicately designed DNA molecules are nested in the heart of each cell, where they act as blueprints for the creation of RNA molecules, a slightly different form of nucleic acid which carries the genetic information from the nucleus to the site of protein synthesis. The RNA molecules serve in turn as templates, guiding the formation of specific proteins that are needed to carry on the life of the organism.

These special protein molecules are enzymes (a name given to any catalyst that occurs in living things.) An enzyme encourages and accelerates one specific chemical reaction. Even though it may undergo change while the reaction is occurring, it is re-formed at the end and is ready to perform the same expediting function in an endless stream of identical reactions, providing a shortcut along which the process can be led again and again. All living things make use of enzymes to control the complicated reactions that take place at the cellular level. Thus at every step the intricate chain of command is sustained and augmented by the remarkable tendency of matter to facilitate the creation of other highly organized forms of matter. The information from the DNA molecules passes through this chain, directing the shape and development of each living thing. It ordains the five-fold symmetry of the sea urchin, the brilliant blue color of the angelfish, the towering form of the giant redwood tree.

Several decades ago most biologists believed that life originated with the creation of a single self-replicating molecule. This possibility, although still favored by some, is now generally considered to be too simplistic. There is a world of difference between a single nucleic acid molecule and the simplest living organisms we know today, bacteria and blue-green algae. These organisms consist of single cells of the most primitive kind without nuclei. But they have two key characteristics: the ability to metabolize and the ability to reproduce themselves. The cell metabolizes by taking in substances from its environment. It breaks down the compounds, rearranges the atoms into molecules needed for its own maintenance, and rejects those that are not useful. To avoid losing the essential chemicals the cell must be separated from the rest of the environment by some kind of boundary surface. How and when did the cell membrane come into existence?

Scientists studying this problem (notably Alexander Oparin in Russia and Sidney Fox in the United States) have discovered that there is a tendency for concentrated water solutions of polymers, when heated, to separate spontaneously into droplets. A selective process causes the concentration of polymers to be greater inside the droplets than it is in the surrounding medium. These droplets, known as "coacervates," are set off from their surroundings by a kind of membrane, a thickening of the polymer molecules around the outside of the droplets. This natural process has both a selective and a protective aspect; it creates and preserves the identity of the whole.

In some cases these coacervates grow in size as more polymers are absorbed from the medium and reactions take place within the droplet. Unwanted inorganic materials are diffused through the membrane and rejected as waste. Thus a very simple kind of metabolism takes place even in this unit of nonliving matter.

When the drops have grown to a certain size they break up into several daughter droplets. Those that receive the proper complement of protein molecules can continue to grow, but the supply of these chemicals becomes gradually diluted and eventually the process cannot sustain itself. If there were a mechanism for ensuring that each daughter droplet was endowed with the pattern for manufacturing the essential molecules using raw materials from the environment, then we would be witnessing a primitive form of life.

It should be noted that the experiments reported here used polymers which are not primitive; thus the activities of these coacervates can only be considered as analogues of the natural processes that may have taken place in the primordial seas of the earth. However, if we accept this analogue as a guide to the processes that led to the origin of life, we see that two lines of chemical evolution were involved. There was the line that led to the formation of protein molecules—under the right conditions these molecules might have drawn together and formed a protective membrane that set them apart and created a new whole. The other line culminated in the formation of the nucleic acid molecule which—given the proper environment—could reproduce itself endlessly, and would provide a pattern for the reproduction of protein molecules.

The generally accepted theory of the origin of life assumes that these two evolutionary lines of development took place within a single unit of matter. But there is another possibility. Perhaps the two different lines of chemical change evolved in two separate units, producing a self-replicating molecule and something like a coacervate. Then at a moment in the early life of the planet the two met and fused. Each unit carried special talents and they pooled their resources to create a new and larger whole, an organism with potentialities far exceeding any that had been known before.

Fusion processes are ubiquitous throughout nature, as we have seen. Quarks fuse to make nucleons; nucleons fuse to make larger and more complex atoms; atoms fuse to make molecules and crystals. In the next chapter we will see how certain primitive organisms fused to create higher forms of life. Fusion provides a mechanism for rapid evolution, compressing the slow process of random mutational events winnowed by natural selection. The average waiting time required for a number of favorable mutations to occur at a single site is much greater than if the same number were divided between two units whose evolution took place in parallel.

In fusion events matter plays an active role; it shows a propensity to associate with other complementary forms of matter. This active quality is manifested in inorganic as well as in organic matter, but in life it is more dramatically displayed. "I like to think," remarked Cyril Ponnamperuma, head of the Laboratory of Chemical Evolution at the University of Maryland, "that the universe is in the business of making life. There is a characteristic of life in almost everything, and at a certain point it becomes recognizable."

But what is this certain point? Or, to put the question another way, what is life? In fact, no precise dividing line between living and nonliving matter has been identified. The ability to replicate, to pass on organized information, is perhaps the most generally accepted criterion. But even this definition does not make a clear distinction.

A single gene would theoretically be able to reproduce itself if it existed in a medium that contained all the essential building blocks. But inorganic matter can also reproduce itself under the right circumstances. We have just seen how coacervates divide and grow and divide again. A star in the process of formation may rotate so rapidly that it breaks up into two parts which then become reorganized in the shape of spheres, thus creating two smaller stars. Given the proper interstellar medium, these stars attract more matter and can grow by accretion to the size of the original star. Or take the example of an ice crystal falling through a cloud. It grows as it attracts water vapor molecules, and when it becomes too large it shatters. Each piece then attracts more water vapor molecules, restores (at least to some degree) the symmetry of its crystal structure. It grows and eventually it, too, shatters. Thus the original crystal replicates itself millions of times over, and precipitates a snowstorm.

Suppose we narrow the definition of life and require that replication be carried out with strands of nucleic acid. Even with this restriction the distinction cannot be clearly drawn. Viruses represent a whole class of organisms on the dividing line between living and nonliving matter.

Viruses come in many sizes and complexities, but none of them can grow or reproduce independently. They do so only within a living cell. Viruses contain nucleic acid, a double strand of DNA or RNA, and a protein shell which serves to isolate and protect the nucleic acid and also provides a mechanism for invading a living cell. In the bacteriophage (a virus that attacks bacteria), for example, the shell is shaped like a tiny tadpole with a polyhedral head and a stubby tail. When this virus encounters a bacterium the tail is strongly attracted and clings to the cell. A digestive enzyme at the tip of the tail then catalyzes the dissolution of the cell wall with which it is in contact. Through this opening the nucleic acid enters the cell; the protein coat, having served its function, remains outside. The invading DNA now takes over the cell chemistry, overruling the genetic instructions of the bacterium. The protein materials within the cell are reorganized to produce virus protein and virus nucleic acid. Finally the bacterial cell is sucked dry and falls into shreds. In its place are hundreds of bacteriophage viruses, each one complete with its deadly nucleic acid coiled within its protein shell.

When the viruses are separated from the bacteria they have invaded they exhibit none of the characteristics of life. They do not grow or metabolize or reproduce; they cannot be cultured in a test tube. But they can be crystallized and kept in this inert state for many years.

Is the virus alive or isn't it? Perhaps it should be classified as something that is not alive when it exists alone but becomes alive when it combines with the protein molecules encased within a cell. This union might be considered an analogue of the event that may have happened long ago when a nucleic acid molecule fused with a droplet of protein. It is an analogue, not a duplication, of this historic event because the virus must have as its partner a living cell.

A similar synthesis of the basic ingredients of life—the union of the sperm and the egg—created each one of us. Like the virus, the sperm is very small with a rounded head and a tail. It contains a single strand of nucleic acid and enough protein to form its casing, which isolates and protects the genetic strand and also facilitates the penetration of the membrane of the egg. The much larger egg contains many proteins in solution, enzymes, and other ingredients essential to the mechanisms of metabolism and reproduction. It also contains a single strand of nucleic acid. When this fuses with the strand carried by the sperm, a complete set of genetic information for making the finished organism is created by the union.

The egg by itself is not alive and neither is the sperm but life results from their fusion. Because they are both products of living matter we cannot say that in the act of conception we are witnessing the creation of life *de novo* from nonliving matter. However, there are interesting and suggestive similarities in the two cases we have cited, the meeting of the virus and the cell, and that of the sperm and the egg. In both of these events we witness the spontaneous joining of nucleic acids and protein molecules within a protective cell membrane. The result is a new and more capable unit of matter. This new organism absorbs substances from its environment and rearranges them for its own purposes. It grows and divides and replicates itself, thus extending its occupancy of both space and time.

We have found this same tendency to create larger, more versatile, and longer lasting organizations of matter far down the scale of simpler and simpler things. The fact that we cannot find a sharp dividing line between life and nonlife suggests an underlying unity—that both aspects of reality are part of a single creative process. And life is just a stage in the organization of matter.

Life is a stage in the organization of matter.

Plate 7
Feather (Photograph © 1985 by Andreas Feininger).

Plate 8
Frost Flower (Photograph © 1985 by Andreas Feininger).

In so saying, we have pushed the mystery back to the very heart of the cosmos at the moment of its birth. Loren Eiseley put it this way:

> Men talk much of matter and energy, of the struggle for existence that molds the shape of life. These things exist, it is true; but more delicate, elusive, quicker than fins in water, is that mysterious principle known as organization, which leaves all other mysteries concerned with life stale and insignificant by comparison. For that without organization life does not persist is obvious. Yet, this organization itself is not strictly the product of life, nor of selection. Like some dark and passing shadow within matter, it cups out the eyes' small windows or spaces the notes of a meadow lark's song in the interior of a mottled egg. That principle—I am beginning to suspect—was there before the living in the deeps of water. . . .
>
> If "dead" matter has reared up this curious landscape of fiddling crickets, song sparrows, and wondering men, it must be plain even to the most devoted materialist that the matter of which he speaks contains amazing, if not dreadful powers, and may not impossibly be, as Hardy has suggested, "but one mask of many worn by the Great Face behind."

FROM AMOEBA TO MAN

*Evolution does not mark out a solitary route. . . .
it takes directions without aiming at ends, and it
remains inventive even in its adaptations.*

HENRI BERGSON
Creative Evolution

For at least a billion years after life first appeared, the earth was
a harsh and inhospitable place, almost unbelievably different
from the planet we know today. Deadly rays of solar energy
shone through a poisonous atmosphere onto the bare rocks of
the continents. Winds, laden with fumes and dust from active
volcanoes, blew with unimpeded force across the lava fields and
the dark face of the ancient seas. Far beneath this wave-tossed
surface, in the semi-gloom where only dim blue light penetrated,
the phenomenon of life was beginning in a very tentative and
seemingly insignificant way to take hold. In spite of the great
complexity of form embodied in these earliest single-celled organisms, they must have been very small indeed. Judging from
the average size of the simplest life forms today—the bacteria—

25,000 of them laid side by side would not have measured half an inch. But at first there was probably only one . . . then two . . . then four. . . . From such infinitesimal beginnings, step by step life grew, adapted to its demanding environment, burgeoned into a million diverse and sturdy forms. But equally important, it gradually molded the environment to suit its own needs.

Life altered the atmosphere and gentled the sunlight. It turned the naked rocks of the continents into friable soil and clothed them with a richly variegated mantle of green which captured the energy of our own star for the use of living things on earth, and it softened the force of the winds. In the seas life built great reefs that broke the impact of storm-driven waves. It sifted and piled up shining beaches along the shores. Working with amazing strength and endurance life transformed an ugly and barren landscape into a benign and beautiful place where wildflowers carpet the hillsides and birds embroider the air with song. The story of how all this came to pass is a shining example of the creative forces working within nature.

In the atmosphere of the infant earth there was no free oxygen and no protective layer of ozone (the molecular form of oxygen which contains three atoms instead of two). The presence of ozone acts as a screen, filtering out the ultraviolet radiation from the sun. These rays are so energetic that they would have damaged the delicate molecules of living matter, and even have broken up the large organic molecules floating near the surface of the sea. The earliest life forms must, therefore, have lived in the half-lighted depths where water shielded them from the ultraviolet rays.

These simple living things subsisted on the organic compounds which had been synthesized by sunlight or by lightning. They absorbed these molecules through their cell membranes, broke them down with the help of their enzymes, and reorganized the atoms into molecular forms suitable for their own use. But as the living cells divided and multiplied, the supply of nutrients in the ocean was gradually depleted. And the diminishing supply probably limited the population of the first life forms.

Sometime very early in the history of life a supremely important new ability was acquired. A special protein molecule which we now call chlorophyll enabled some of the living cells to use the energy of sunlight to manufacture their own food supply from carbon dioxide, water, and mineral compounds that were abundant in the environment. The first true plant had been created.

This event, almost as important as the appearance of the first living organism, happened at least 3.5 billion years ago. Fossils of that age in the Warrawoona Formation of western Australia carry the imprint of primitive unicellular photosynthetic organisms similar to the blue-green algae of today.

The fossil record is so fragmentary that gaps of billions of years tantalize the scientists who are attempting to follow the winding course of very early evolutionary development. Like the trail left by Hansel and Gretel when the birds had consumed most of the crumbs, only a few scraps of information are left scattered here and there throughout the earth's crust, and many turnings are no longer marked.

After Warrawoona the next clue to the evolutionary trail was also found in Australia. In the Bungle-Bungle Formation a layer of dolomite 1.5 billion years old contains fossils that are believed to be the remains of "eucaryotic" cells. These represent a higher stage of cellular complexity than the cells of the most primitive organisms. The set of circumstances that resulted in the evolution of these elaborately integrated units of living matter is a matter of great interest and some dispute among microbiologists.

Just a few years ago it was assumed that all evolutionary changes occur in small, step-by-step mutations within single units of living matter. The favorable mutations arc selected from the much more frequent unfavorable ones by the process of natural selection. The more efficient organisms compete successfully for the diminishing food supply and reproduce in larger numbers. Thus the favorable mutations are spread rapidly throughout the population and gradually replace the less efficient forms. By its very nature such a process takes place slowly.

A new scientific insight suggests that a more rapid process led to the development of higher levels of cellular complexity. This theory, first suggested by several writers early in this century, has recently been revived, refined, and very effectively supported by Lynn Margulis of Boston University.

According to this model the eucaryotic cell was formed by the symbiotic union of several simpler organisms. This type of cell has an organized nucleus containing paired strands of nucleic acid. In plant eucaryotes the chlorophyll is carried in tiny units— organelles—called chloroplasts. Each one is bounded by a membrane, so it appears to be a little cell within a cell. Other types of organelles are also present in large numbers within the eucaryotic cell. There are mitochondria which, like the chloroplasts, have specialized functions; they facilitate respiration and the processing of food. A third type of organelle are little cilia which make movement possible.

The organelles closely resemble certain types of bacteria. They contain their own DNA and protein-synthesizing machinery. Although they are not capable of duplicating themselves independently, they reproduce themselves within the cell and are transmitted as whole units. The evidence is very strong that these organelles were once free-living organisms that had acquired special attributes: bacteria with whip-like cilia for movement, bacteria with efficient respiratory capabilities, and others that had become photosynthetic. At some point in evolutionary history one or more of these individuals entered into a symbiotic relationship with a host cell. Later others joined the community. As time passed the whole became more and more tightly integrated, the parts more dependent on each other. And from this union evolved the larger cell from which all higher forms of life are built.

The Bungle-Bungle Formation and other rocks around the earth bear witness to the fact that photosynthesis was occurring widely by 1.5 billion years ago. It was altering the chemistry of the earth's crust and the composition of the gases surrounding the planet. The primitive atmosphere consisted principally of four gases: carbon dioxide, ammonia, methane, and water vapor. In the process of photosynthesis the carbon dioxide component was gradually depleted and oxygen took its place. The oxygen combined with ammonia, forming water and nitrogen atoms which came together in pairs to make nitrogen gas. The oxygen also reacted with methane, forming carbon dioxide and more water. The end result was the removal of methane and ammonia from the atmosphere, substituting oxygen, nitrogen gas, and water vapor.

The invention of chlorophyll made an abundant food resource available, but in order to take maximum advantage of it the primitive microorganisms had to live in an environment where sunlight was freely available. It was a fortunate circumstance that the process of photosynthesis created a protective screen so life could exist safely in the shallower, sundrenched layers of the sea. As the amount of oxygen increased in the atmosphere, ozone also began to accumulate, and it filtered out some of the ultraviolet component of sunlight. The populations of microorganisms grew; the ozone layer became more effective; and life was able to move upward again, intercepting more sunlight.

The increasing concentration of oxygen, however, precipitated a new crisis. Oxygen is a powerful chemical that reacts with organic compounds. This characteristic enables it to release the energy stored in sugar and starch molecules, making this fuel available to run the vital body functions. But other reactions with organic compounds make oxygen potentially dangerous. Many of the simplest creatures that existed in the primordial seas would have been killed by exposure to even small concentrations of free oxygen. Very sensitive ones that we can study today cannot tolerate oxygen levels above one percent of the present concentration in the atmosphere. So a protective mechanism was needed in order to guard life forms against oxygen's destructive effects. This mechanism was provided by the mitochondria in the eucaryotic cells. They possess a complex set of enzymes which guide the reactions involving molecular oxygen in ways that make energy available to the organism, releasing it in ways that are not injurious.

No one is sure just when or in exactly what sequence these factors came together in the ancient seas, but it seems apparent that the union which resulted in cells containing mitochondria created a better adapted organism. The microorganisms that did not take part in this evolutionary change continued to exist in old ways, protecting themselves from the dangerous new element—molecular oxygen—by living deep under water, burying themselves in mud or under layers of sediment. Their descendants—the anaerobic bacteria, yeasts, and fungi—are still plentiful today, exploiting these primitive ways of life. Although well adjusted to their particular environment they have played a relatively passive role and have remained virtually unchanged throughout billions of years.

The organisms that found new ways to build higher levels of organization spearheaded the evolutionary process. They experimented with new symbiotic relationships and took advantage of new opportunities as they arose, continuously altering their environment and at the same time adapting to the changes they themselves created.

Many of the single-celled life forms discovered another way of exploiting the advantages of cooperative behavior. The algae cells lived in colonies; vast numbers grew together, making thick mats of vegetation that were arranged in layers like a head of cabbage. From the examples of colonial species that we see in nature today—the honey bees, the termites, the coral reefs—we know that many benefits derive from such collective interaction. In a colony each organism metabolizes and reproduces as a separate unit, but the individuals converge and coordinate their activities in ways that contribute to their own survival as well as the survival of the species. The principle involved in convergence and fusion is the same; it is only a matter of degree.

One of the special advantages of the more highly organized eucaryotic cell is the possibility of sexual reproduction. Most unicellular organisms reproduce by binary fission. The cell divides, going through an elaborate, carefully orchestrated maneuver whose end result is the creation of two identical daughter cells. Each one is endowed with a copy of the double strand of chromosomes possessed by the mother cell.

There is, however, another type of cell division which results in "haploid" cells, each of which carries only one set of the chromosomes of the parent cell. In some ways this reduced inheritance is a disadvantage; the cells are not able to replicate themselves as efficiently as in the manner of binary fission. But nature has turned the new circumstance to a creative advantage. When one of the haploid cells meets a similar cell—perhaps from a different ancestral stock—they fuse and form a new individual whose genetic makeup is unlike any that had existed before. Thus greater variety is achieved, although at the cost of a somewhat slower process. Several simple cell divisions could have occurred in the time required to effect the production of a single individual carrying two different sets of chromosomes. Nevertheless, the long-term result of sexual reproduction is to speed the process of evolutionary development, because the probabilities for favorable change are greatly enhanced. Imagine, for example, that in an asexually reproducing species ten mutations occur in a given time. When those ten are put to the test of natural selection the probability of any one being favorable is very small. On the other hand, in a sexually reproducing species the recombination of ten different mutations can produce more than a thousand variants. There is a much greater chance that an improved type will emerge and be selected.

Before we have left the subject of the primitive beginnings of sex let us take a moment to observe several of the related experiments that nature has carried out. All of these illustrate the same tendency toward unification, the coalescence and melding of single organisms.

Through a microscope single-celled animals, the simplest flagellates, for example, can be seen to multiply rapidly by binary fission. But occasionally two individuals can be observed to come together and melt completely into one. The individuals taking part in this conjugation often come from different, not very closely related genetic stock. Conjugation is a much rarer event than normal fission and, again, it represents a delay in the reproductive process. However, a totally new individual has been created.

A variation on this process is seen in another kind of primitive organism, the ciliates. Under the microscope we can see two individuals come to lie side by side and then exchange bits of their nuclei. This act of union does not involve the whole cell, as in the previous example, but the exchange of genetic material produces a similar result, bringing into being a uniquely endowed individual.

These experiments in evolutionary progress were undoubtedly carried out in the ancient seas, but the fossil evidence is too fragmentary to provide a coherent record of the very early periods of earth history. The tiny one-celled forms of life had no shells or bony structures to act as favorable materials for fossilization. Only occasional imprints of their soft microscopic-sized bodies are left to guide us, and there is almost no record of the long period between 1.3 billion and 900 million years ago. By the end of that time a very rich assemblage of microorganisms had evolved, as can be seen at the Bitter Springs Formation in the North Territory of Australia. Here fossils of fifty-six species of algae, fungi, and other plants have been identified. Such rapid evolution indicates that sexual reproduction had probably been taking place.

Another long break in the record occurs between Bitter Springs and 700 million years ago. This gap is particularly tantalizing because during these millennia another very important milestone in the history of life was reached. The first multicellular plants (green algae) made their appearance. Their fossilized imprints have been found in both Norway and Australia.

It is a remarkable and still unexplained fact that almost three billion years passed between the appearance of unicellular colonies and the earliest recorded multicellular organism. Although we do not know the details of how the step from unicellular to multicellular life was taken, it is apparent that the same form-building tendency which generated the many fusion and convergence processes that preceded it is manifested again in the production of this new form of matter, so pregnant with potential for the future.

Did this evolution occur by a slow process of mutation, as the individual cells in a colony of single-celled organisms sacrificed very gradually some of their autonomy and a more tightly organized system was achieved? Or was the change brought about by a rare event? Perhaps two cells in the process of binary fission found ways of interchanging chemical messages that served to coordinate their activities and bind them together into a single organism. If we could follow this evolutionary change in detail it would provide more insight into nature's ways of achieving greater complexity of form.

Colonies that assume the shape of long chains may have been the ancestors of the first true multicellular plants. Many of the primitive seaweeds and fungi are built of such chains. The cells attach themselves end to end and thus construct a long filament of single cells. By elaborate interweaving of these filaments quite solid-looking structures can be built. Mushrooms and toadstools, for example, are formed of networks of strands. This type of structure, however, does not carry through to the higher plant forms. Beginning with the mosses there is no trace of filament design; the single units are joined together like cells in a honeycomb.

The union of separate individuals opened up possibilities that proved to be almost limitless. Within a multicellular organism regiments of cells divide the labor and perform specific functions for the organism as a whole. It is the duty of some cells to anchor the plant on solid rock, of others to build and maintain the supporting structure, of others to collect and process the food. A small number of specialized cells are charged with the responsibility of reproducing the organism and extending the life of this newly created living unit.

However, in spite of the intense specialization and apparent subjugation of the individual cell to the needs of the larger whole, each cell still retains its own separate potentialities. In an emergency it can take on an entirely different role. It can even maintain its own existence in isolation, given a favorable environment with the proper nutrients. Like the electrons in an atom, the molecules in a crystal, the amino acids in the protein chains, the units involved in this new level of matter do not relinquish their individual identity. Building of form has taken place at each stage without the loss of the ones before.

As new possibilities were created by multicellular organization, life responded, creating a diverse and whimsical assortment of soft-bodied things. Not only plants but animal forms evolved, feeding on the abundant vegetation and thus indirectly living on the energy of sunlight. Great colonies of algae constructed undersea gardens that provided a favorable habitat for other living things.

About 570 million years ago shelled aquatic life suddenly made its appearance. The earliest shelled fauna were mysteriously unlike any that inhabited the sea in subsequent eras. Within a very short time, however, these died out and other species similar to our snails and oysters and sea urchins took their places. Fauna with hard skeletons and a bone protecting a central nerve cord evolved about 500 million years ago. The most primitive fish had heavy bony armored plates around their heads, and instead of movable jaws they had suckers with which they grubbed for food along the sea bottom. The fossilized remains of shelled species began to leave a detailed record of life's history in the rocks of the earth. We can observe how the number of marine species increased in a spectacular way. Cephalopods, whose living relatives include the octopus and squid, left large numbers of distinctive funnel-shaped shells. They grew to frightening size, sometimes attaining more than twenty feet in length, and seem to have dominated the shallow intercontinental seas that were very widespread at that time. Rugose coral evolved and flourished in these warm waters. Smaller colonial organisms—bryozoa and graptolites—also built reef structures. Very extensive beds of these fossils have been found throughout many continental areas today.

While this evolution was altering the forms of individual organisms, large-scale changes brought about by the whole network of living things were proceeding at an ever accelerating pace. As life multiplied, the amount of free oxygen in the air increased and the density of the ozone layer continued to build. By 400 million years ago it had reached a level of concentration similar to that which exists in the atmosphere today. It provided an adequate shield from ultraviolet radiation, so organisms no longer needed the protection of water. The stage was now set for the next important step in the evolution of life, a step that must have taken place on the wild, beautiful borderland between land and sea.

• • •

Of all the enchanting places on earth, those that lie on the threshold of the sea are the most deeply charged with vitality and power. My favorite one is a wide sand beach that runs the length of Harbour Island in the Bahamas. It is a stretch of coral sand tinged just the faintest shade of pink; so in bright sunlight the high strand is glistening white, but where the shadows lie at the base of the dunes there are ripples of pink, and along the water's edge where the waves continually caress its surface the sand is coral-colored, gleaming like the smooth underside of a great conch shell, and edged with whorls of white foam left in little windrows by each retreating wave.

The best time to walk this beach is just before low tide, when the broad swath all along the sea has been swept clean of weed. The sand is firm and cool underfoot and as I walk I can watch the great swells rolling in from the sea. One after another they come, rising high as they approach the shore in long graceful arcs. Just before they break the light shines through them, transparent, aqua-colored, like Venetian glass.

Gradually, as the tide falls the reefs that had been submerged just offshore begin to catch the waves, impaling them on gnarled brown fingers. And all the colors of the sea intensify—azure, ultramarine—there are no words to describe the depth and brilliance of the blue.

Now at the edge of the beach a small reef formation is partially exposed. It is a fascinating labyrinth of little interconnecting pools surrounded by ramparts of porous limestone rock made years ago, perhaps when the sea level was higher. But the reef is long since dead, unable to survive the twice daily exposure to the extremes of aerial existence.

Life is severely challenged in this intertidal zone. With every ebb tide it is bared to the harsh realities of life on the land where temperature differences are much greater than in the sea. The heat of the sun and the drying breath of the wind bring the constant danger of desiccation. Gravity is a more troublesome force than that experienced in the sea where it is counteracted by the buoyancy of the water medium. On land it makes all movement and transport of nutrients more difficult.

At slack tide, looking down through the crystal water of the little coral grottos I can see at least half a dozen different forms of marine life. Each in its own individual way has dealt with the dangers peculiar to this threshold of the sea. Barnacles grow in profusion along the walls of the coral rock. They have attached themselves with a cement so strong that they can only be pried loose with a sharp-edged tool. Their rounded streamlined shapes offer minimum resistance to the waves. Little limpets cling to the rocks with suction cups. They protect themselves from the impact of falling water with fluted shells shaped like old-fashioned parasols. The perimeter of each shell forms a firm contact with the rock and prevents desiccation of the soft body lodged within. A sponge, one of the most primitive of sea creatures, has adopted a low profile, flattening its body into a thin yellow crust along the most sheltered surface of the coral. Here, too, an ivory-colored sea urchin is nestled above the reach of the water at the lowest tide. And several small dark shells, embedded so tightly in the rock that they appear to have grown there, reveal the presence of feather dusters that have retracted their tentacles. In these undistinguished curls of limestone it is difficult to recognize the beautiful sea flowers they will become when the tide returns, like fringed chrysanthemums with petals that wave gracefully as they gather nutrients from the sea.

In and out among the tidal pools with their clinging long-term residents move several species of miniature fish; little brown ones dart and hide beneath the rocks, and tiny pale-blue ones weave threads of silver through the quiet water. But soon the tide will rise. The waves will come sweeping back, breaking over the reef with crushing force. All the little creatures that have survived the long dry hours will then be subjected to the great stress of the turbulent sea.

Here on the borderland between ocean and land the drama of life has played out some of its most important scenes. Here the rehearsals took place in which marine organisms learned the arts of life on land—rehearsals relentlessly repeated twice daily as the moon drew the waters of the earth back and forth in an endless rhythm across the shining beaches and the rocky shores.

We can follow the logic of the remarkable evolution that transformed soft streamlined sea creatures into organisms tough and resilient enough to leave the sheltering cradle of the water medium and conquer the continents. Selective pressures were great in the demanding and stressful environment of the intertidal zone. Plants developed strong body structures that protected them from the force of the waves. They evolved ways of gripping securely onto the rocks or the soil so they would not be washed away. They invented methods of retarding desiccation of their cells, and ways of transporting moisture from one part of their bodies to another. All of these new talents stood life in good stead when it took on the challenge of colonizing the land.

Still there is a mystery not totally illuminated by the doctrine of life passively buffeted by wind and sea, slowly altered by chance mutations, and winnowed by natural selection. If these forces were the only ones at work then easy environments where few changes were demanded in order to achieve maximum survival of the species would have been favored above the more difficult ones where more mutations and adjustments were needed. But the enormous success of the adaption to land cannot be ignored. Although only 30 percent of our planet's surface is land, almost 80 percent of all species are terrestrial, and these include the most highly evolved forms. We know that life does not follow the path of least resistance. There is something within life, within nonliving matter, too, that is not passive—a nisus, a striving that is stimulated by challenge. Steadily throughout geologic time life has moved out from easy to difficult environments and now has populated almost every nook and cranny of the globe, from the frozen wastes of Antarctica to the boiling sulfur-laden vents at the bottom of the sea.

The first land plants that left their imprints in rocks 400 million years old were photosynthetic, but instead of leaves they had green stems pierced with air holes. They had evolved a system of moving water from the base to the upper branches. Stiff thorny trunks provided a strong body structure and root-hairs gripped the soil. These plants with their minimal equipment for terrestrial existence were the pioneers that opened up a whole new way of life.

But there was one more problem that had to be solved before

adaptation to aerial life on a broad scale could be achieved. The element nitrogen is a minor constituent of living matter, composing only about 1 percent. But none of the necessary proteins or nucleic acids can be built without it. Nitrogen is plentiful in the earth's atmosphere (78 percent is elementary gaseous nitrogen). However, this abundant source cannot be used directly by multicellular plants or animals. Nitrogen in the gaseous form has two atoms in each molecule, and this arrangement is so stable that very large amounts of energy are required to break the molecule apart and free the individual atoms, enabling them to enter into other associations. This process is called "fixing" nitrogen.

On the primitive earth nitrogen was available in usable form. Ammonia (a compound of nitrogen and hydrogen) was a major component of the atmosphere, and it was also dissolved in the sea. Then, as we have seen, the process of photosynthesis began to produce free oxygen, which reacted with ammonia, forming water and releasing gaseous nitrogen. By the time oxygen had reached close to its present level and the protective layer of ozone had become concentrated enough for life to exist on land, ammonia was no longer present in any appreciable amounts. At that time lightning may have been the main source of fixed nitrogen. A flash of lightning can heat the air to 10,000°C in a fraction of a second. In this searing heat nitrogen atoms are torn apart, and later they combine with oxygen and water to form nitrates, which are dissolved in raindrops, distributed on the soil, and taken up by the plants. This source of fixed nitrogen was not sufficient, however, to supply the needs of all the life forms that began to take over the earth.

At a moment early in the evolution of life on land two organisms joined together and formed a symbiotic relationship which has continued to this day and which has greatly increased the amount of fixed nitrogen available for the nourishment of living things. What lightning accomplishes by brute force is achieved in silence and usually in total darkness by minute and extremely primitive organisms living along the roots of a special type of land plant. Certain anaerobic bacteria that have never learned how to manufacture their own food supply or how to live safely in the presence of oxygen have this one extraordinary talent. When they are present in the nodules of legume roots they secrete an enzyme called "nitrogenase" which fixes nitrogen in a way that bypasses the need for large surges of energy. It accelerates the fixation of nitrogen without itself being used up in the process. And it does this work so efficiently that today it provides the largest natural source of available nitrogen in the world, although the total amount of nitrogenase is estimated at less than ten pounds. It is an interesting fact that these bacteria, living alone, are not capable of fixing nitrogen. The chemistry of the plant is needed to stimulate the production of nitrogenase. Thus the combination of plant and bacteria produces an effect that neither one by itself could achieve. This symbiotic relationship solved an important environmental problem for all life and made possible extensive populating of the continents.

Within a relatively brief moment of geologic time the descendants of the first land forms evolved, radiated, and multiplied. Not only did they deal effectively with the problems of life on land, they gradually altered the land to nurture and sustain life in general more abundantly. The root systems of the plants helped to break up the hard rocks and aerate the surface layers. Their leaves shaded the ground from direct sunlight, slowing evaporation and reducing temperature extremes. Before many eons had passed a tide of green vegetation had flowed across the lowlands, following the watercourses and the marshy borderlands of the sea.

The first animals left the gentle sheltering water environment about 390 million years ago. They were spider-like creatures that had developed the ability to breathe air. Perhaps this talent had evolved in shallow bays on the shores of a retreating sea or lake. Thirty or forty million years later a fish that had acquired this same skill crawled out onto dry land. It had internal nostrils, so it could breathe with its mouth closed. And it had another helpful attribute, too. It had bones in its fins, so it could "walk" on land. These structural changes had actually turned this fish into the first amphibian.

And yet even with these advantages the first land animal entered a harsh and frightening world. Every step must have been a painful experience as the force of gravity pressed its soft body down upon hard ground and its fins dug into sharp stones. The search for food demanded constant movement and effort; nourishment did not just float by, as it had in the sea. The battle to survive in this dangerous new environment called forth reserves of strength and endurance that had been untapped in the less demanding medium of the tropical waters.

But why did these pioneer organisms battle to survive? We know that all living things have a drive for self-preservation, but we have taken this phenomenon so much for granted that it does not seem to need explanation. Perhaps we should take a fresh look at this strong motivation which causes each individual to put forth a tremendous effort to prolong its extension in time. In fact even inanimate organisms tend to preserve their identity and resist dissolution. Self-preservation is a universal phenomenon. Without this mysterious force there would have been no step-by-step building of more advanced levels of form. Each step must be retained in order for the next one to rise upon it.

We do not have the space in this book to follow in detail the path of evolutionary change from amoeba to fish, to amphibian, to mammal, to primate, and finally to man. It is a devious way, with many dead ends and blind alleys. Most biologists are wary of assigning any direction to this process. They hesitate to call one form more "advanced" than the ones from which it evolved. They point out that some primitive organisms, like the anaerobic bacteria, are as well adapted to their environment as human beings are to theirs. But this fact does not disprove the reality of progress, as Julian Huxley has argued:

> It is . . . an empirical fact that evolutionary progress can only be measured by the upper level reached, for the lower levels are also retained. [The retention of lower levels] has on numerous occasions been used as an argument against the existence of anything which can properly be called progress; but its employment in this connection is fallacious. It is on a par with saying that the invention of the automobile does not represent an advance, because horse-drawn vehicles remain more convenient for certain purposes, or pack animals for certain localities. . . . we

can discern a direction—the line of evolutionary progress: increase of control, increase of independence, increase of internal co-ordination; increase of knowledge, of means for co-ordinating knowledge, of elaborateness and intensity of feeling—those are trends of the most general order.

The classic argument about the reality of evolutionary progress can be seen as a problem in semantics. If one defines progress as the degree of improvement in adaptation to the environment then some of the very simplest organisms meet this test as well as more complex ones. But if one assumes with Huxley that evolutionary progress is measured by improvement in internal organization, in greater control, independence, and potential, then the changes that have occurred throughout time have produced spectacular advances as complexity has increased from the simple one-celled organism without nucleus or organelles to the human being containing as many as a hundred trillion eucaryotic cells.

The word *evolution* comes from the Latin *evolvere,* meaning to roll out, to unfold. The etymology is unfortunate because it suggests a development predetermined from the first moment of creation and revealing itself in time, like the opening of a bud. This is a pretty analogy but it does not match the facts. Although several schools of biological thought have espoused this idea, a realistic look at the development of life refutes the concept. Theodosius Dobzhansky described some of the objections, pointing out that evolution has taken a long time because it was not a simple matter of unfolding or uncovering something that was there from the beginning. It has involved many false starts, and failures leading to extinction. There have also been great breakthroughs of creative novelty. "If the universe was designed to advance toward some state of absolute beauty and goodness, the design was incredibly faulty."

On the other hand a process that depends upon trial and error inevitably spins off imperfections and failures. The result of each change cannot be foreseen; it can only be tested and rejected if the new form is not favorable. This produces a mixture of good and evil, harmony and discord, such as we see in the world around us. The fittest are those that survive best, but many intermediate cases also exist. Clinging tenaciously to life, they occupy small specialized segments of the environment.

The creativeness of evolution is apparent everywhere. Like an imaginative artist experimenting with different ways of building form, sensitive to color and line, pattern and shape, it has tossed off one original design after another: the deep golden cup of the honeysuckle blossom, the iridescent splendor of the peacock's tail, the cool, precise melody of the whipporwill's song. But errors are evident, too—the mongoloid child, the hunchback, the three-legged calf. They are not as common as one would expect if chance alone were responsible for change; but the existence of tragic mistakes would not be expected at all if a pattern, already created, were simply being unrolled.

The drama of the evolutionary development that led to man can be followed in a vastly accelerated and somewhat fragmentary version within the human womb. The fertilized egg—a single-celled organism—begins to divide and for several weeks the embryo is similar in shape and organization to the simplest multicellular creatures, indistinguishable also from the very early embryos of many other classes of animal—lizards, fish, small birds, or mammals. At the end of the fourth week the embryo has begun to develop a distinctive body form—but quite a curious one. The most prominent feature is a large muscular tail which curls forward and upward, rivaling the trunk in massiveness. It looks for all the world like the broad powerful tail of a fish. The heart, main arteries, and neck region are also built on the same plan as those of a fish. There are four pairs of clefts on the sides of the throat—clefts that obviously correspond to the gill slits of a fish. As development proceeds, the fish-like characteristics are gradually modified and by the end of the second month the embryo begins to take on a generalized mammal-like form. The seven-month human embryo is strikingly similar in certain respects to a chimpanzee or a gorilla. The soles of the feet are turned toward each other like the palms of the hands—a trait that is useful in the other primates for grasping the branches of trees. At this age the human embryo has thick hair on the head, eyebrows, and lips, just like the ape embryo of the same age. Finally, in the last two months the distinctively human characteristics are acquired.

Although this recapitulation—as it has been called—does not reproduce exactly all the earlier stages of development, fragments of the process are retained like an imperfect memory of the evolutionary path. In some mysterious way command instructions for ancestral forms are still preserved in the genetic code of the more highly developed forms of life. New levels are achieved without totally destroying the ones that went before.

A number of other intriguing and still unexplained facts are known about the long journey that led from the first living thing to man. There is considerable evidence today that the ladder of evolutionary progress was not entirely composed of an immense number of very small steps. The fossil record shows long periods of stability punctuated by sudden bursts of change. Although some paleontologists attribute this pattern to an incomplete record, others believe that leaps in evolution did take place, and they are attempting to understand the process in this new light.

The changes that transformed living protoplasm have occurred within the tiny coils of DNA concealed inside the nucleus of the cell. The complexity of this minute packet of matter defies imagination. A single gene is constructed of at least a hundred thousand atoms; each chromosome contains somewhere between two thousand and twenty thousand genes; each species carries a characteristic number of chromosomes—from two to several hundred. The deciphering of this fantastically elaborate code is far from complete, but spectacular progress has been made in the last few decades. Geneticists can count the chromosomes, stained and illuminated in high-resolution microscopes. They can see the inversions and crossovers, and can even estimate the number of mutations needed to produce the chimpanzee from the same stock that spun off the orangutan.

In a recently published scientific study the chromosomes of man, chimpanzee, gorilla, and orangutan were photographed side by side. The similarities are very impressive. Eighteen of twenty-three pairs of chromosomes are identical in the four species. Of the remaining pairs, four show only slight variations. The differences between orangutan and man are the most pronounced, between gorilla and man slightly less; the chimpanzee and man are almost identical. Only the chromosome pair known as No. 2 demonstrates a major variation between man and the other three species. The human chromosome No. 2 appears to be the result of a fusion of two chromosomes which are separate strands in the chimp, the gorilla, and the orangutan. In this way the number of pairs was reduced from twenty-four to twenty-three, an event that was unique and has not been reversed. This may have been the alteration that produced the large evolutionary step from primate to man.

The implications of this scientific fact are exciting and humbling. Such a relatively minor juggling of the microscopic units of matter inside the nucleus of a single cell has set loose forces that have shaken the world. It produced a new form of living matter that has altered the surface of this planet, its atmosphere, its waters, and its community of living things. It led to the invention of writing, music, and the arts. Reaching out beyond the confines of its own planet, this novel life form has explored the surfaces of Mars and Venus, has sent messages to the farthest reaches of the galaxy, and has shaken the foundation blocks of matter on which the whole elaborate tower of creation has been built.

What awe-inspiring powers and what grave dangers might be realized by rearranging just another letter or two in this alphabet of life?

BEING AND BECOMING

Thou canst not stir a flower
Without troubling of a star.

FRANCIS THOMPSON
The Mistress of Vision

The concept of a universe in a state of genesis requires a level of sophistication that mankind has only recently attained. A human life is so brief; it is like a tiny candle flickering in the vastness of cosmic time—snuffed out before the world has evolved enough for the changes to be apparent: before the oceans have moved halfway around the globe, before a new pole star has replaced Polaris in the northern sky, before the mountains have been returned grain by grain to the sea. But through language and writing, the ability to pass thought down from generation to generation has extended the little flame of thought at least a thousandfold. Mankind has also learned to play with the dimension of duration—to alter its rate of flow and thereby gain an intuitive understanding of the relativity of time.

The invention of the moving picture was an important step in building this new perception. Take a series of still photographs or drawings, each one exactly like the last except for an almost imperceptible change in one direction. Run these stills rapidly past the eye. Action suddenly springs forth and a story unfolds—the beautiful maiden flees from the villain; the hero, roused by her cries of distress, gallops to the rescue. Movement is seen and instantly understood even though each of the frames is static, timeless, and unchanging.

Time-lapse photography carried the conceptual understanding of time a little further. Snapshots of a thing known to be in a state of becoming can be timed to accelerate the process, so a rosebud grows and blooms before our eyes, then withers and drops its petals. A whole growth cycle encompassing perhaps a dozen days is compressed into the span of a minute or two. Then becoming stands forth as the essential reality; states of being exist only as stills snipped from the moving picture of continuous change. Conversely, very rapid action can be slowed down or stopped by stroboscopic photography and the instantaneous shape of things in action is revealed. By these devices we can reset nature's clock to fit the tempo of our own lives.

If we could drink a magic potion and slow down the process of our own life cycle—perhaps a hundred-thousand-fold, so we would live in the rhythm described in the Old Testament: "a thousand years . . . would be as yesterday when it is passed and as a watch in the night." Then cosmic evolution would unfold before our eyes like the opening of the rose. We would see stars being born while others flamed and died. We would see them gather together and circle in great whirlpools of matter; we would see how the galaxies cluster together even as space is expanding between them. Suddenly the immense cosmic distances would become accessible. We could travel to Sirius and back before the "night watch" was out. With probes traveling at the speed of light we could circumnavigate the Milky Way and explore its halo of hydrogen and dust. Seen from this new perspective we would view with pity the old life-measure of three-score years and ten, ephemeral and insignificant as the lifetime of a mayfly that is hatched and mates and dies in a few hours of a summer afternoon.

In imagination we can play thus with time. Einstein might have called the exercise a thought experiment—useful because it illuminates the whole cosmic process of change more vividly to our mind's eye. Imagination is the magic potion that enables us to rise above the limitations imposed by the great differences in scale between cosmos and man.

Looking in this way at the universal process of evolving Form we see that the myriad states which appear to be static are really changing, moving—though almost imperceptibly—in one direction. Everything is in a state of transition to a more complex, more highly integrated system. The familiar varieties of matter—molecules, organisms, and species—are not really fixed states of being, they are stages of becoming, like stills snipped from the continuous process of transformation. And there are also many intermediate stages which display the properties of wholeness. Organized from within, they act in ways that increase their extension in space and time. They preserve and regenerate themselves, and they take advantage of opportunities to participate in more advanced levels of form. In some cases it is possible to identify a series of transitional states which we can imagine run together in a moving picture to reveal how the higher levels of organization may have come gradually into being.

In the last chapter we touched on the question of how multicellular life developed from simple unicellular forms. We found that single-celled organisms like the blue-green algae had adopted a system of colonial organization very early in the earth's history, but that several billion years passed before true multicellular organisms evolved—or at least before they left proof of their existence in the rocks. The passage of this great length of time suggests an extensive period of experimentation with intermediate methods of organization. In fact we can see in the world today that there is a large middle ground occupied by living things that are neither wholly unicellular nor wholly multicellular.

Many algae, protozoa, fungi, and bacteria are colonial for a portion of their life cycles. Some of these come together for a period of time as loose clusters of cells; the whole unit exhibits no internal structure or differentiation of parts. In other cases a primitive level of integration is attained; for example, some of the cells of the colony become specialized for feeding, others for reproduction, and so on.

The cellular mold known as *Dictyostelium discoideum* is an interesting case in point. It starts life as a single amoeba-like cell, living in the soil. It feeds on bacteria, grows, and reproduces by binary fission. When the food supply is depleted, growth stops, and then the single cells begin to converge, all moving radially toward a central location where they construct a conical mound of cells. When all the nearby cells have been collected, the mound falls over and becomes something like a small fat slug. Worm-like it moves over the surface in the direction of light. If a new food supply is encountered on this migration the slug may disintegrate and the individual cells disperse to lead individual lives again. Otherwise it enters into an elaborately organized reproductive phase. It constructs a long stalk enclosed in a sheath of cellulose which terminates in a basal disc at the bottom and a round "head" containing a mass of spores at the top. Eventually the spores are cast off and begin life again as single cells. The stalk and basal disc falls apart and the cells die. In making this reproductive structure the original amoeba-like cells differentiate into three distinct cellular types (spore, stalk, and basal disc) which are quite dissimilar in size, shape, and function. Experiments have shown that the role each cell will play is determined by the order in which it enters the aggregate. Those that arrive first become the leading portion of the migrating slug and later they form the base of the stalk. Cells that arrive next at the collecting point become upper-stalk cells and spores. The last to arrive form the tail of the slug and the supporting basal disc. The differentiation is controlled by chemical signals passing between the cells and organizing the unit. This process begins while the cells are still separated from each other. They are attracted and caused to aggregate by specific chemical agents produced at the center. But even after becoming part of the larger unit the individual cell still retains its own genetic constitution intact. If separated out it can reproduce by binary fission. Its progeny is capable of passing through the same cycle, converging, forming slugs and stalks that bear spores. As few as ten cells can provide the components and the chemical interactions needed to carry out this reproductive phase.

The sponge provides us with another striking example that lies on the borderline between cellular colonies and primitive animals. The sponge passes most of its life as an individual organism whose basic plan consists of a tubular structure with many tiny pores. Water is drawn in through the pores, passed along a series of canals, and then expelled through a vent. The water movement is controlled by thousands of minute "collar cells" that line the passages and flail threadlike whips vigorously in the desired direction of flow. As the stream of water passes, nutrients carried in it are absorbed by the cells. In addition to the collar cells there are special cells that compose the skin, the middle layer, and the skeleton on which the soft body is supported. Most sponges can reproduce themselves by regeneration. If pieces are broken off each one grows and reforms a single, functioning sponge.

Almost a century ago an imaginative biologist named H. V. Wilson wondered how far the sponge could be broken down and still retain the power to remake itself. Cutting it up into small pieces or pulling it apart with needles did not destroy this power, so Wilson put the organism through an even more drastic process of disintegration. He placed a small amount of sponge tissue in a cloth bag. Then, using a pair of forceps, he squeezed the contents gently through the cloth under water. The mesh was so fine that the sponge tissue was broken down into single cells or very small assemblies of cells. The fragments fell through the water, making a cloud of finely divided material that was collected on a glass plate at the bottom of the vessel. The single cells, now freed from their association, began to look and behave like amoebae. They extended little lobes and moved along the glass surface. But before long a remarkable process began to take place. Whenever two single cells approached each other they extended filaments and touched; then they promptly united into a single body. A third cell was quickly added, then another and another, making a small mass. Separate nearby masses united, producing large colonies and eventually one single assembly that formed a crust on the surface of the slide. In the space of a few hours or days the aggregate had regenerated itself, and then differentiation began, producing the four different varieties of sponge cell. The tubular structure was built, the skeleton, the middle body, and the skin.

Biologist Edmund Sinnott, commenting on this experiment, said: "In some way the structure of the whole sponge, with its specific pattern and its cellular variety, is immanent in each individual cell. How this is possible is a secret still locked in the innermost structure of protoplasm. Evidently each tiny living unit is a much more complex thing than it appears to be, for it bears within itself an image of the whole organism. . . ."

It is significant that the larger structure is the preferred state. When released from the sponge organism, the tiny individual cells return as quickly as possible to the more complexly integrated system.

The trend toward higher degrees of organization can also be traced through various stages of colonial behavior in insect societies, such as ants, termites, and bees. Here, too, we find single organisms, then loose assemblies, then increasingly higher and more differentiated systems. The history of evolutionary changes of this kind can be inferred from the varying degrees of social organization displayed in bee communities today.

The bee is a relative newcomer among the species that inhabit the earth. Just a short time ago, geologically speaking, when dinosaurs roamed the earth and shallow seas covered much of the North American continent, there were no true bees and not a single blooming plant added fragrance or color to the lush jungles of green vegetation. Bees and flowers evolved together, making their first appearance about a hundred million years ago, and their fates have continued to be entwined ever since. The bee serves a necessary function in fertilizing the flower, and the flower provides the pollen and nectar that nourish the bee.

We usually associate bees with a highly organized hive society, so it is surprising to discover that among the more than 10,000 species of bees living today, 95 percent are solitary insects, only meeting with others of their kind long enough to mate. They do not share the work of gathering food or caring for the young. The most primitive of these species is the *Hylaeus,* an almost hairless insect with only a rudimentary tongue and no pollen baskets on its hind legs like those of the honeybee. The *Hylaeus* lives from hand to mouth, so to speak. It swallows the pollen, transporting just enough to fill a few crude nest cells that it has hollowed out of plant stems. In each of these it deposits one egg. Then the life cycle of the bee is complete. It does not survive the frosts of autumn and never sees its offspring, which—if all goes well—emerge the following summer.

The first step toward cooperative lifestyle is exemplified by the mason bees, so called because they cement together fragments of stone with their own saliva to fashion the cells that serve as nurseries for their young. Instead of building their nests in isolation, they form communities with others of their kind. On a single rooftop there may be thousands of these bees working in close proximity, although each one builds its own individual nest, fills it with food and eggs, and seals it up. Then, their personal responsibilities discharged, they join in a cooperative activity for the general good of the colony. Together they lay a protective coating of mortar over the whole collection of cells.

The *Halictidae* have evolved a more specialized social system. These bees nest in the earth or in rotten wood, burrowing long tunnels sometimes as much as six feet below ground level. Hundreds of underground cells are dug out close together and lined with a strong paperlike substance. Like the mason bees, each *Halictida* constructs its own individual cell. However, all the bees enter and leave the nest through a common entrance, and this hole is guarded by a single female bee. Under normal circumstances she uses her head to block the opening, and when members of her community return to the nest laden with pollen, she backs down from the entrance to allow them to pass. But if danger threatens she reverses her position and presents her sting at the opening of the tunnel. If the guard is killed or removed another bee quickly takes her place. Females of this species that survive the winter lay eggs in the spring and these hatch into sterile worker bees ready to nurse the young that emerge later in the season. By dividing the labor of providing for the next generation and guarding the nest, the *Halictidae* improve the average chances of their own survival and of their species as a whole.

The bumblebee community is an even more highly organized society. In temperate zones the life of the colony begins with the emergence from hibernation of a fertile female. Until she raises her first brood this queen's existence is as lonely as that of the solitary bees. In the early spring she hunts for a suitable nesting site, often picking an abandoned mouse's nest. She cleans it out, then makes an egg cell and a tiny thimble-shaped honeypot out of wax produced by special glands along the segments of her abdomen. When she has collected enough pollen to construct a soft bed in the wax cell, she lays her eggs and sits on the cell like a brooding hen. The eggs hatch into grubs, spin cocoons, and finally emerge as sterile worker bees. These become the queen's corps of willing helpers, gathering food, building more egg cells, and caring for the next batch of young bees. By the height of the summer season the queen is surrounded by hundreds of her offspring, all cooperating for the good of the community. At the end of the season when autumn frosts are threatening, the last eggs are laid and these develop into fertile females and drones. After they have mated, the fertilized females find some protected place to hibernate for the winter, while all the other members of the society are killed by the cold. In tropical climates, however, bumblebee colonies survive for many years, dividing their populations occasionally by swarming when conditions become too crowded.

The honeybee society is the apex—at least the present apex—of the progressive evolution of more and more highly structured bee communities. In an average honeybee hive there are about 30,000 workers, 500 to 2000 drones, and one queen. By cooperative action the honeybees meet the challenge of surviving the winter. They store up hoards of golden honey, providing enough food to last through the months when no flowers bloom. And they use a simple but effective method of combatting the winter cold. Bees, like all other insects, are cold-blooded. They can only raise their body temperature by muscular activity. When cold weather strikes, the bees cluster in the hive, flap their wings vigorously and move backward and forward in a seemingly endless jig. The tempo of the dance slows down or speeds up as the outside temperature rises or falls. Within the cluster there is a continual transfer of individual bees from the periphery toward the center and out again. They spell each other, alternately resting and dancing. Thus the work, the hardship, and the advantages of warmth are shared.

The honeybee colony has also solved another survival problem. If an accident overtakes the bumblebee queen in midseason, that colony is doomed to extinction, but the death of a honeybee queen poses no such threat. In response to chemical signals a substance called royal jelly is manufactured in the heads of the worker bees. This jelly, fed to larvae, produces a new queen and the life of the hive goes on almost uninterrupted.

Throughout the levels of organization of bee colonies we find increasing powers of regeneration—an ability that is present to some extent in all organisms, both living and nonliving. The lizard, the starfish, the ice crystal, even the tiny hydrogen atom, are able to restore themselves to their original form after an external force has disrupted them. This ability becomes more sophisticated as higher levels of organization are attained.

Sponges and bee societies are examples of higher orders built from the union of like organisms, but many unified systems are also built on the cooperation of different species. These are known as symbiotic relationships, and we have already discussed a number of examples: the bee and the flower, the legumes and the nitrogen-fixing bacteria, perhaps also the chloroplasts and the mitochondria that united with the very primitive unicellular life forms. In symbiosis as in colonial behavior a series of stages can be recognized, starting with loose cooperative arrangements and moving through various degrees of mutual interdependence.

Some combinations are so close that the two participants spend their entire lives together. The lichen, for example, which was believed for many years to be a single organism, is really a double plant created by the symbiosis of mold-like fungi and single-celled green algae. The algae are photosynthetic and can live alone, but when they are housed within the mass of fungi threads, they live a more sheltered and favorable existence. The fungi, on the other hand, are dependent upon the photosynthetic ability of the green algae cells to manufacture food. They can exist independently only in a medium rich in the necessary nutrients. The symbiotic relationship of these two has produced a plant that is strikingly different from its separated partners in both appearance and biochemistry. Lichens secrete compounds—medicines, dyes, perfumes, and poisons—which cannot be made by either component alone. Lichens are plant pioneers, creating soil on bare cliffs and volcanic lava flows and coral islets. They break down the rocks chemically, converting the minerals into biologically useful compounds. Gradually they turn the hard surfaces into soft, fertile soil. Lichens are so successful that they have been able to colonize an unusually wide range of environments. They are found almost everywhere on the land surfaces of the earth, from deserts to tundra to rain forests. In the polar regions they are the most plentiful plant, covering the otherwise bare rocks along the coasts of Antarctica and Alaska. They are found far above the snowline; yet they also grow luxuriantly in the tropics.

Although the genetic code of both partners is not passed on as a single unit, the methods of reproduction assure that in most cases the next generation will receive its endowment of both fungus and alga cells. The lichen has three ways of perpetuating itself. The first is simple fragmentation. When pieces of the lichen are broken off, these portions, containing both fungi and algae, grow and regenerate, making new plants. The lichens also produce buds which are liberated from the plant and carried on the wind to new locations. Each bud contains threads of fungus and at least one alga cell. In the third method, however, the fungus produces spores which do not have any algae associated with them. If these spores settle and germinate in a location where algae are not also present they soon die of starvation.

Is the lichen one organism or two? This example as well as many other closely cooperative relationships points up the need to recognize many different levels of internally organized systems. When we think of the living world in terms of discrete forms and species, we are focusing on certain stages of the transformation process. The many intermediate systems appear to be anomalies that do not fit conveniently into any one category. But when we think in terms of continuous change moving toward the formation of more complex self-regulated wholes, the intermediate states fall into place as essential parts of the process.

The coral reefs that border many of the tropical land areas on earth provide an excellent illustration of the many different gradations in the evolution of Form. An entire spectrum of living systems can be observed there, ranging from the simplest partnerships to the most complex communities.

No one should miss having glimpsed at least once in their lifetime the fantastically beautiful and elaborately interrelated world of the coral reef. I was past middle age when I saw it for the first time. Because I am not a strong swimmer and suffer slightly from claustrophobia when I try to swim under water, I had never ventured to put on a snorkel mask and look beneath the shining face of the sea. But then one autumn I spent a vacation on the tropical island of Bora Bora, which is encircled by one of the most wonderful reefs in the world. To observe this reef divers had traveled for thousands of miles, bringing with them their scuba and snorkel equipment. Their descriptions of this remarkable underwater world stirred my imagination. So one day I put on a mask to take a look at a little coral head that grew in relatively shallow water a few hundred feet from the beach.

As my mask broke the mirrored surface I entered a veritable wonderland, like Alice stepping through the looking glass. Here nature was dressed in its most gala attire, a dazzling display of elaborate shape and vibrant color—deep-violet mushroom coral and moss-green brain coral, bright orange and red fire sponges. There were delicate pastels, too—pale daisy coral like tiny flowers and lavender anemones and staghorn coral tipped with blue. Swaying in the gentle rhythm of the waves, lacy purple fans and bronze crinoid plumes moved in graceful unison as in a dance. Little ripples on the surface of the sea deflected the sunlight, creating a vivacious pattern of light and shadow that passed con-

Countless examples speak of an underlying force respon-
sible for designing a world of order and beauty.

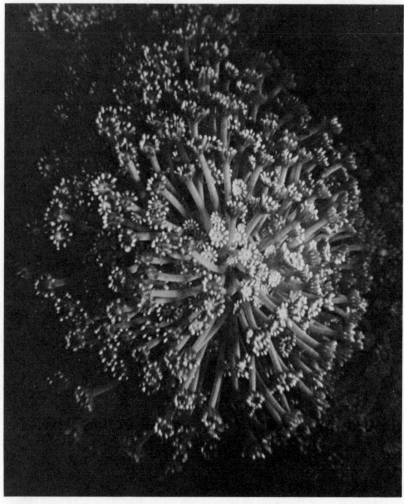

Plate 9
Daisy Coral Formation (Photograph by Douglas Faulkner).

Plate 10
Queen Anne's Lace (Photograph © 1985 by Andreas Feininger).

tinuously back and forth across the reef, reminding me of the romantic light patterns cast by the many-faceted mirrored globes that animated the ballroom scenes of long ago.

A school of miniature gold-colored fish swam in the crystal-clear water above the reef—hundreds of them, no bigger than fireflies. But they moved in perfect formation, wheeling and turning as a single body. A shaft of sunshine reflected on their shining bodies, creating a single glittering surface of light. Directly below me, in a narrow cavern between two pillars of coral, I saw a flash of brilliant red. As I watched, a flame angelfish swam slowly out of its hiding place. Turning with regal grace, it displayed its gorgeous scarlet dress, edged with royal blue and accented by five bands of midnight blue—one of the most striking designs in nature.

Since that first experience I have visited this underwater wonderland countless times. And each occasion reveals something new. The reef seems to hold an unlimited store of treasures. To the casual observer it appears to be a dazzling assembly of many beautiful individual things all competing for food and space and solar energy. But actually this galaxy of creatures exemplifies every known type of cooperative relationship between living things.

The reef structure is built by millions of tiny coral polyps, a simple animal consisting of just two layers of cells. Because it is animal, the polyp cannot synthesize its own food supply. It must depend upon the organic material floating in the sea or nutrients manufactured by plant life. The tropical waters are very nearly transparent and this clarity is largely due to the fact that they contain relatively few microorganisms compared with the rich profusion present in the oceans of higher latitudes. If coral polyps had to depend solely on the supply of plankton they would not thrive and could never have built the hundreds of thousands of square miles of coral formations that now exist beneath the surface of the sea.

Sometime in the remote past the coral polyp formed an active partnership—a symbiotic union—with single-celled photosynthetic algae of the type known as zooxanthellae. These algae capture the energy of sunlight and manufacture the nutrients needed by the coral polyps. The waste materials excreted by the polyps provide a rich source of food for the algae, and the reef structure furnishes them with a stable, protected environment. This mutually beneficial relationship works to maximum advantage when the living portion of the reef is near the ocean surface, so the zooxanthellae receive a considerable amount of sunshine.

The coral polyp is the architect and builder in the partnership. It extracts calcium from the sea, and combines it with carbon dioxide and water to make the hard, resistant material from which it fashions the tiny pentagonal cell in which it lives.

Genesis of new life occurs most frequently by budding. A tiny polyp forms on the surface of the parent cell. It grows until it attains a comparable size and then it too divides. The algae also multiply by binary fission and are simply passed along from one coral polyp to another.

In addition to asexual budding, the coral reproduces sexually. At certain times of year sperm and eggs develop within a fold of the polyp's internal skin. Released when ripe, the sperm float across the living surfaces of the reef and are drawn into the mouths of polyps containing mature eggs. The fertilized eggs produce larvae that are endowed with microscopic hairs for propulsion through the water. Millions of these larvae—or planulae, as they are called—are released from the reef at times which appear to be related in some mysterious way to the phases of the moon. Off the coast of Australia drifts of reddish-brown planulae stream from the Great Barrier Reef at the season of the full moon in winter. But in summer this breeding activity coincides with the new moon.

The coral planulae are better equipped to deal with life challenges than the lichen spores we considered earlier because the young polyps carry along with them a starter population of their partners, the zooxanthellae. Thus the symbiotic relationship is preserved even in this process of sexual reproduction, and those planulae that find a favorable location in which to settle are immediately provided with a source of food—their own private vegetable gardens. As the polyps grow and bud they build a new reef structure characteristic of the species to which they belong. Identical units are added cell by cell and remain attached to each other, in a manner very similar to the way honeycombs are made or the way crystals grow.

There are hundreds of coral species and these take on a wide variety of forms, from the tightly packed masses of brain coral to the delicate branches of finger coral. In given environments certain shapes are more favorable than others. The open formations provide more surface and allow better penetration of sunlight. The compact, massive shapes are most able to resist rough surf and surge conditions.

Working patiently through the sunlit hours, year after year, the coral polyp creates an underwater castle containing countless individual rooms and hiding places for the diverse forms of marine life that inhabit it. Within its sheltering nooks sponges, seaworms, and anemones take up residence. The soft corals find the structure an ideal platform on which to grow, waving their graceful plumes and delicate fans in sunlit waters beyond the reach of the rough surf of the open sea.

Among the inhabitants of the reefs many symbiotic relationships exist. There are the little cleaner fish and shrimp that work diligently for many hours picking irritating parasites off large fish. This convenient source of food seems to be considered fair exchange for the cleaning service, so the cleaners are allowed to live safely within easy reach of the large predatory jaws. A diver described how this métier is pursued by the Pederson shrimp, a tiny organism prevalent in Bahamian waters:

> When a fish approaches, the shrimp will whip its long antennae and sway its body back and forth. If the fish is interested, it will swim directly to the shrimp and stop an inch or two away. The fish usually presents its head or gill cover for cleaning, but if it is bothered by something out of the ordinary such as an injury near its tail, it presents itself tail first. The shrimp swims or crawls forward, climbs aboard and walks rapidly over the fish, checking irregularities, tugging at parasites with its claws and cleaning injured areas. The fish remains almost motionless during this inspection and allows the shrimp to make minor incisions in order to get at subcutaneous parasites. As the shrimp approaches the gill covers, the fish opens

each one in turn and allows the shrimp to enter and forage among the gills. The shrimp is even permitted to enter and leave the fish's mouth cavity. Local fishes quickly learn the location of these shrimp. They line up or crowd around for their turn and often wait to be cleaned.

When not actively engaged in this work the shrimp spends its time in close proximity to a sea anemone, clinging to it or occupying the same hole. In this position it is protected from other marine organisms by the anemone's poisonous tentacles. The anemone seems to recognize the shrimp as a valuable associate, allowing it to live unmolested in a situation that would be fatal to many other species. For its part the shrimp serves as a lure, attracting fish within range of the anemone's tentacles.

These cooperative arrangements all act as units within a much larger and more elaborate system. The living reef itself functions like a giant organism; it metabolizes, grows, and reproduces itself. The limestone-like structure provides the skeleton that supports the body of the reef. Its many little passages and chambers are the pores through which sea water is drawn and circulated to all the "cells" of the coral body. Useful minerals and organic matter are removed from the water and transformed into the various nutrients needed for the life of the reef.

The little individual organisms contribute in a wide variety of ways to the successful functioning of the whole. Sponges burrow into the limestone structure, increasing its porosity. By pumping the water through their own bodies, they speed the circulation of this "life-blood." Vast numbers of fish feed on the plankton in the sea water, and as they swim over the reef they drop a shower of fecal pellets that are consumed by the coral polyps, excreted, and passed on in turn to the zooxanthellae. These pellets are rich in nitrates and phosphates, elements essential to the manufacture of organic matter by photosynthesis. The algae also secrete a limy deposit that helps to cement loose fragments and heal minor damages to the skeleton. Torn coral which is vulnerable to invasion by blue-green algae is kept clean by sea urchins, trimmed by butterfly fish and bristlecone worms. Encroachment of one coral species on another is kept in check by a kind of chemical pruning. Coral polyps can extend tiny filaments through holes in the cell walls to absorb and digest the bodies of neighboring polyps. This pruning is a method of maintaining proper balance in the whole system. On a healthy reef no one species is allowed to become rampant; many different species grow side by side in harmony.

Chemical signals circulating through the water provide a rudimentary nervous system for the reef, transmitting information about the food supply, the flow of solar energy, the need for repair of injury, perhaps also the timing of sexual activity. As these messages are received by individual species, they respond in ways that protect, regenerate, and reproduce the larger whole. The reef is an organized system and possibly a superorganism in the process of formation. Participation in this larger whole contributes to the effectiveness of each subunit.

The discovery that everything is related to everything else is one of the most important insights of our time. A century ago Darwin's theory of natural selection focused the attention of philosophers and scientists on the competitive aspects of nature. It placed great emphasis on the struggle for dominance, survival at the expense of others, "nature red in tooth and claw." These are realities of life, it is true; but there is another side to the picture which we are just beginning to understand. A living community is essential to the survival of the species that take part in it. Just as the presence of one organized form of inert matter can act as a catalyst, a template for other units of matter, so the presence of one life form provides protection and nourishment for other living things. Paleontologists have found evidence in the fossil record that communities of diverse organisms have thrived together under favorable environmental conditions throughout the history of life on earth. When conditions changed, the whole interrelated group of living things perished together within a short period of geologic time. Later another community evolved to take its place.

Over and above the living communities, the working partnerships, the symbiotic marriages, we are beginning to recognize the outlines of an even larger organization of matter. It is a system in which we, too, take part, though we never dreamt of its existence until just a few decades ago when we broke free from the force that had bound us for millions of years to the earth's surface. For the first time in human experience we could look back and see our planet whole. Set against the fluid blackness of space, we saw a luminous bubble of matter, variegated with moving patterns of light and shadow, with the delicate shadings of life. It looked as fragile and vital as a diatom floating in the darkness of the night sea.

Since that historic moment we have become aware that the planet Earth may be a single self-organized unit of matter, perhaps even a living thing. This thought was eloquently expressed by Lewis Thomas:

> Viewed from the distance of the moon, the astonishing thing about the earth, catching the breath, is that it is alive. The photographs show the dry, pounded surface of the moon in the foreground, dead as an old bone. Aloft, floating free beneath the moist, gleaming membrane of bright blue sky, is the rising earth, the only exuberant thing in this part of the cosmos. If you could look long enough, you would see the swirling of the great drifts of white cloud, covering and uncovering the half-hidden masses of land. If you had been looking for a very long geological time, you could have seen the continents themselves in motion, drifting apart on their crustal plates, held afloat by the fire beneath. It has the organized, self-contained look of a live creature, full of information, marvelously skilled in handling the sun.

British scientist James Lovelock is another articulate advocate of this new vital interpretation of the earth. In a book published in 1979 he presented a well-supported hypothesis that the entire biosphere together with the atmosphere, the oceans, and the soil forms a self-regulating system maintaining the conditions in which life can flourish. This organism Lovelock calls *Gaia,* after the ancient Greek name for the earth goddess. Evidence for the existence of Gaia can be found in the extraordinary chemical composition of the atmosphere and the oceans. The quantities of the gases present in the air "are improbable by at least 100 orders of magnitude," Lovelock says. And yet this particular combination of elements—extremely favorable for life—has been maintained within very narrow limits over a long period of time.

The same argument, Lovelock maintains, can be applied to the composition of the oceans. A salt concentration over six percent would be lethal for almost all marine organisms. The average saline content today is about 3.4 percent and there is geological evidence that this figure has not varied significantly for at least a hundred million years, although all the rivers have been continually emptying more salt into the sea. These facts suggest that the salt level is under some kind of central cybernetic control.

The uniformity of the earth's climate is another fact that demands explanation. With the exception of a few relatively brief ice ages the temperature range has not changed appreciably since life first appeared on earth over three and a half billion years ago. We know that water in the liquid state existed on the planet at that time, even though astronomical theory tells us that in those early years the energy produced by the sun was at least 30 percent less than it is today.

Various phenomena have been invoked to explain how all these favorable conditions could have occurred by chance and been sustained by a series of coincidences. But as a series of lucky circumstances grows longer it becomes more and more improbable. The same facts can be explained more reasonably as the consequence of a self-regulating system that is actively manipulating the environment and maintaining the conditions most favorable for its own existence. As Lynn Margulis said, "Each species to a greater or lesser degree modifies its environment to optimize its reproduction rate. Gaia follows from this by being the sum total of all species connected, for the production of gases, food, and waste removal, however circuitously, to all the others."

The fabric of living things has adapted to changing conditions eon after eon, while at the same time continuously altering the environment to suit its needs. Even after great extinctions damaged large portions of this delicate web of life, it regenerated and flowered anew. Like the many other wholes that we have observed throughout nature, Gaia has taken an active part in maintaining and extending its own existence. It is not just passively pushed around by external forces. In all these characteristics we recognize an integrated whole—perhaps even a living organism. We have found that life is a stage in the organization of matter; the point where an organism becomes a living thing is simply a matter of definition. Forms that are not living today may cross that line tomorrow, since all organisms are constantly evolving—becoming more complex, more tightly integrated, and more capable of increasing their occupancy of space and time.

Beyond our own planetary system how many higher levels of organization are in the state of becoming? A galaxy shows characteristics of self-regulation which we can understand best by looking at galaxies other than our own. Viewed from afar, the form of Andromeda can be grasped at a single glance. But the Milky Way, seen from inside, is just a hazy track of light across the night sky. Only with the help of higher mathematics and giant computers can its total shape and patterns of motion be understood.

If the universe itself is an organism with properties that transcend those of its parts, we may never know its nature. We can never hope to step outside it, look back, and see it whole. The best we can do is to study its parts and put them together in our imagination, reasoning by analogy with organisms that we have seen from both inside and out.

Lovelock found evidence of Gaia in the improbable constitution of the earth's air and sea. Such high degrees of improbability maintained over long periods of time suggested the presence of a self-organized system. This insight comes to mind when we hear that cosmologists are experiencing difficulty in imagining the exact balance of expanding and contracting forces that could have supported a stable and surprisingly homogeneous universe that has not collapsed from gravitational contraction nor diffused into infinity throughout fifteen billion years. British chemist P. W. Atkins observed that if the fundamental forces had been slightly different life and mind could not have evolved. "The balance of the strengths of these forces is critical to the emergence of conscious life," he said. "If nuclei were bound slightly more weakly, or slightly more strongly, the universe would lack a chemistry; and life . . . would be absent. If the electric force were slightly stronger than it is, evolution would not reach organisms before the sun went out. If it were only slightly less, stars would not have planets, and life would be unknown."

Many large-scale phenomena are difficult to explain using the traditional concepts of force and energy—the collapse of a dust cloud to make a star, the creation of a galaxy or a cluster of galaxies. In fact, improbable arrangements of matter and energy are common throughout the universe: pulsars that rotate at 500 times a second, stars packed so tightly that no light can escape from their black centers. And here on earth we find a most unlikely collection of smaller things: a diamond crystal with its precisely ordered pattern of a thousand trillion atoms all exactly alike, a buttercup whose golden blossoms have been shaped on the same DNA template from generation to generation for a hundred million years, a single human being whose form in space-time involves the cooperative activity of a hundred trillion cells.

Throughout the cosmos we find evidence of order at many different levels, all interrelated and all in the process of becoming. As John Muir remarked, "When we try to pick out anything by itself, we find it hitched to everything else in the universe." The cautious methods of science may require many more generations to untangle the threads of this intricately woven web. But poets sometimes have a way of striking to the heart of things which reason must approach by slow and circuitous routes. Francis Thompson, writing at the end of the nineteenth century—before the space age, before the invention of computers, before the awakening of ecology—said:

> *Thou canst not stir a flower*
> *Without troubling of a star.*

CHAPTER 6

ENTROPY AND EVOLUTION

One is constantly reminded of the infinite lavishness of Nature. No particle of her material is wasted or worn out. It is eternally flowing from use to use, beauty to yet higher beauty.

JOHN MUIR
Gentle Wilderness

In 1851 William Thomson, who later became Lord Kelvin, presented to the Royal Society of Edinburgh a paper on the dynamic theory of heat. This paper set forth the principle of conservation of energy, later known as the First Law of Thermodynamics, and another principle concerned with the dissipation of useful energy, now called the Second Law. Kelvin, who held the chair of natural philosophy at the University of Glasgow, worked at a time when improvements in mechanical devices such as the steam engine were transforming life in the Western world. Along with Nicolas Carnot in France and James Prescott Joule in England, he laid the foundations for the science of heat and its transforma-

117

tion into mechanical work. These contributions had a deep and wide-ranging influence on the physical science of the nineteenth century.

In several later papers Kelvin used his theories of heat conduction to calculate the age of the sun and earth. Assuming that the sun is an incandescent liquid mass which receives no heat from without, Kelvin showed that it must have been cooling since its formation, and that it probably has illuminated the earth for not more than 100 million years—certainly not more than 500 million. The earth, too, as most geologists believed, must have been much hotter in the past. Given the present rate of increase of heat with depth below the surface, the planet must have solidified from a molten state sometime between 20 million and 400 million years ago, the most likely figure being about 100 million.

These figures caused Charles Darwin great concern because his theory of natural selection required a very long period of geologic time. In a hundred million years the slow process of survival of the fittest could not have produced the great variety of species found on the earth today.

Charles Lyell was also distressed by Kelvin's conclusion. Lyell's interpretation of earth processes demanded an extended history of stable and basically uniform conditions. But Kelvin's arguments were very persuasive and convinced many of his contemporaries because they were based upon a precise mathematical calculation. Lyell and Darwin, on the other hand, were not able to support their belief with figures; biology and geology were not exact sciences. History has proved them right, however. Although Kelvin's mathematics was correct, his assumptions about the melting temperature of rocks under pressure were later proved to be erroneous. Furthermore, Kelvin did not take into consideration heat generated by thermonuclear reactions and radioactivity within the earth.

The law of conservation of energy became one of the cornerstones of physics and stood unchallenged for more than half a century. Then in 1912 Albert Einstein theorized that matter and energy are different forms of the same thing—a thesis that was dramatically confirmed by the atom bomb. Kelvin's First Law was replaced with a broader conceptual scheme, the conservation of matter-energy.

It may seem surprising that well-established "laws" of science are later shown to be untrue or only partial truths. But scientific laws are only well-established theories, and as such are always on trial. It has regularly happened in science that even the greatest theories are modified by subsequent work, and are often found to be special cases of a wider generalization. Newton's law of gravitation was supplanted in 1919 by Einstein's theory of relativity, a broader synthesis which embraces Newton's law as a first approximation. Scientific theories are working hypotheses rather than creeds. The concept of an ultimate truth, essential to religious faith, is useful in science only as a horizon toward which we proceed and which is forever retreating before us.

Kelvin's Second Law of Thermodynamics is more complicated than the First, and more difficult to explain to the nonscientist. Stated as simply as possible, the Second Law says that every spontaneous change is accompanied by an increase in the randomness of the energy distribution within the system. The quantity which measures this randomness or disorder is called *entropy,* from the Greek word meaning a turning or change. This quantity was especially interesting to Kelvin and his fellow physicists because they were concerned with heat energy for operating steam engines and other mechanical devices. Heat, you remember, is the average kinetic energy of very small particles moving in a random way. Differences in temperature provide motive power that can be translated into work, and large temperature differences can be translated with greater efficiency than small ones. Therefore, as heat energy becomes more evenly distributed, less mechanical work can be gotten out of the system. This loss of differentiation in the distribution of heat represents an increase in entropy.

Many illustrations of the loss of useful energy predicted by the Second Law can be found. A waterfall, for example, can be harnessed to turn a wheel and do mechanical work like grinding grain or spinning wool, but in the process some of the energy is converted into heat of friction and is "wasted."

The Second Law also applies to chemical and nuclear reactions. When conditions are right these happen spontaneously, and in each case some of the energy that has entered into the reaction is rejected as heat or radiation. Therefore, entropy has increased, randomness and disorder (at least in this limited sense) have increased. In these and all other ways involving the transfer of heat and energy the Second Law has proved to be infallible. We can easily understand why scientists and even philosophers began to draw ever more sweeping conclusions from this useful theory. If "disorder" is increasing with every spontaneous change, then with the elapse of time disorder must be increasing in the cosmos. The universe is like an engine running out of steam, becoming less and less differentiated, more random, as time goes by. The trend postulated by the Second Law is the exact opposite of the one proposed in this book. This important difference must be examined with care.

The reason for the apparent conflict between these two theories lies in the definition of disorder. As measured by entropy, disorder is a limited concept, principally concerned with the distribution of random motion. It does not take into consideration the fact that natural form-building activities are occurring throughout time. Self-integrated units of matter exhibit powers which oppose the trend to randomness and produce orderly events.

In the man-made world shape and arrangement are frequently imposed upon systems from the outside. Such systems differ in a fundamental way from self-organized units. A deck of cards arranged in proper sequence of numbers and suits is an example of imposed order. The arrangement is lost as the deck is used, and effort must be put into restoring it. A bridge is built of steel and concrete with the expenditure of effort. The finished structure is the result of imposed forces, and unless work is constantly put into its maintenance the bridge will deteriorate with time. The hydrogen atom, however—the ice crystal, the protein molecule, the coral polyp—are units whose forms are organized from within. They protect themselves from dissolution and repair themselves when they have been damaged.

The physical sciences in general do not recognize this distinction. All orderly arrangements are treated as though they were imposed upon the system. It is perhaps only natural that the earliest attempts to explain the world were based on principles derived from human experience. In order to build a shelter, to mold a clay pot, to fashion a pair of boots, a formal arrangement is impressed on passive materials. The assumption has been made that all form is imposed on brute matter because this is the way mankind creates organization. This assumption attempts to explain the world in the way human beings can manipulate it. Nuclear and electromagnetic forces are held responsible for impressing order upon the atom, the molecule, and the crystal. Gravity imposes order on the planet, the stars, and the galaxies. Although biological systems do appear to operate under a different set of laws, life is considered a special case—a field apart.

On the other hand, I have postulated a continuum where no sharp dividing lines exist from the littlest quark to the galaxies and the human brain—one cosmic process of becoming. All organisms, both living and nonliving, act in ways that increase Form in space and time. These activities occur spontaneously in several different ways: (1) by *synthesis* of smaller into larger units; (2) by *selection*, facilitating the creation of new organisms; and (3) by *self-preservation* and regeneration, thus increasing their average life span. Taking each of these formative processes in turn we will analyze several examples.

Consider first the order arising from *synthesis*—for example, the spontaneous reactions that create one helium atom from four hydrogen atoms. This important process is taking place in most of the stars throughout the cosmos. If we add up the amount of matter in four hydrogen atoms we find that it is slightly more than the amount used to build the helium atom. This leftover matter, less than one percent, is converted into energy and released as heat and radiation. As interpreted by the Second Law, the wasted energy is less useful energy; so randomness and disorder have increased as a result of this fusion process.

But what about the more complex assembly of matter that has been created? The helium atom is a much more elaborately structured unit of matter than the hydrogen atoms of which it was made. It is a whole that is more than the sum of its parts, possessing new potentials. The helium atom serves as an essential building block for larger atoms. Without helium we would not have oxygen, or carbon, or iron. Without these we would not have amino-acid molecules, or nucleoproteins, or life. The Second Law assigns no increase of order to the higher levels of complexity achieved. According to this law, the vast building activity taking place in the stars represents an increase in cosmic disorder because in the process a tiny percentage of the universal material is "downgraded"—becomes less useful in doing mechanical work for human beings.

This leftover energy spills out into space. It lights the lanterns of a hundred billion stars; it is reflected from our satellite and returns to us as moonlight. It pours in generous warmth from our sun and illuminates our day. And most important of all, this energy builds and nourishes life.

As we have seen, the organic molecules that act as essential units in the living molecule can be synthesized from simpler molecular forms by the infusion of energy. Biologists now believe that on the primitive earth this spark came from lightning and from the most energetic rays of sunlight. Thus the energy that appeared to be wasted in the creation of helium atoms probably took part in the synthesis of much higher orders of complexity—amino acids and proteins. These molecules later came together to form the first living things. Feeding upon other molecules synthesized by sunlight, these organisms reproduced themselves and multiplied their orderly arrangement in space and time. Soon they learned to use the energy of sunlight directly through photosynthesis; so solar energy continued to sustain their existence.

Biologists, aware of these facts, have occasionally raised questions about the validity of the Second Law in defining the direction of all spontaneous change. Living matter obviously evades the decay into equilibrium, and it produces higher states of differentiation. This apparent contradiction of the Second Law has been "explained" by the physicists, who point out that in measuring changes in entropy one must consider a totally closed system. Order may increase in one place if a comparable decrease in order is sustained in another part of the system. The order produced by living things, they maintain, is more than offset by the increase in entropy represented by the downgraded energy produced by fusion processes in the sun. Since the outflow of solar radiation is very large and the phenomenon of life relatively small, all the orderly arrangements produced by living things can be accounted for without challenging the conclusions drawn from the Second Law. In the universe as a whole, radiation from the stars is an even vaster source of positive entropy to balance against any small decreases that may occur locally.

There is a fallacy in this explanation, and an important omission. If there were no life in the universe the same outpouring of heat and light from the stars would occur. Life made its appearance spontaneously and has been growing ever since. It has used the energy "downgraded" by thermonuclear reactions in the sun and stars, and at the same time it is producing orderly events and higher degrees of differentiation. Thus at the very least, we can say that the advent of life has reduced the rate at which randomness is increasing in the cosmos.

Furthermore, the energy from the sun was released in the process of creating more complex atoms. This aspect of the change is not counted in the entropy equation. The chain of events goes like this: four simple atoms combine in a series of reactions to make a more highly organized atom plus a small percentage of the total matter-energy emitted as radiation. This leftover energy helps to construct organic molecules which serve as building blocks for life—the most complex self-integrated form of matter known in the universe. The energy from the sun also acts as a nutrient; it has sustained life throughout the billions of years of its evolution on earth. Taking into consideration all of these actions, which occurred spontaneously, we can see that an important increase of order has taken place.

In supporting this conclusion, however, we are in a position comparable to that of Darwin and Lyell confronting Kelvin's pronouncement about the age of the earth. Our arguments are qualitative, while the Second Law is couched in precise mathematical language. At present it is not possible to assign definite numerical values to the different levels of complexity. How much more order is added to the universe by the formation of one helium atom from four hydrogen atoms? Or in the synthesis of amino-acid molecules from their constituent parts? How can we place a value on the creation of the first living thing? Until these processes can be quantified we must be content to point out that the Second Law does not take into account the form-building process of synthesis that is an important aspect of cosmic change.

The second type of creative activity which runs counter to the conclusion of the entropy equation can best be described as *selection*. The Second Law states that differentiation in an isolated system decreases with all spontaneous changes, and this has been shown to be true in every situation involving the irregular thermal motion of atoms and molecules. Entropy reaches a maximum when the positions and velocities of the molecules and atoms are distributed completely at random. A state of equilibrium then exists. For example, suppose there are two flasks filled with gases and connected by a narrow tube. The gas in one flask is heated up while the other remains at room temperature. When the source of heat is removed, the molecules gradually move from one flask to the other until the energy is evenly distributed. The temperature differences have disappeared; order, in the sense of a high degree of differentiation, has decreased and randomness has increased as predicted by the Second Law.

The British physicist James Clerk Maxwell, who was active at approximately the same time as Kelvin and was strongly influenced by his work, invented an imaginary scenario to dramatize the principle of increasing randomness. He postulated a tiny being placed in the narrow tube between the two flasks and capable of directing the flow of traffic. The "Little Demon" could preserve or create differences in temperature by allowing only high-speed molecules to move from left to right and only slow-speed molecules from right to left. This selective activity would reverse the direction of entropy change predicted by the Second Law. But there are no Little Demons in nature, Maxwell said. The particles in all mixtures of gases and fluids become more and more evenly distributed with the elapse of time. Spontaneous natural processes go in the direction of increasing randomness.

As a matter of fact, however, we know that there are some situations where selection does take place in a way that is very reminiscent of Maxwell's Little Demon. Take a glass of water and dissolve in it half a cup of pure sugar. The sugar molecules spread out until the sugar and water are thoroughly mixed. Where we originally had sugar and water, we now have sugared water, and entropy has increased. If the solution is allowed to sit uncovered, the water will evaporate, leaving a more concentrated sugar solution. Water molecules have passed from the liquid to the more disorderly gaseous state with the absorption of energy; so in this case, too, entropy has increased. But when the concentration of the solution is sufficiently strong, a little sugar crystal begins to form and to grow by selecting from the solution those molecules that match its own. With a rapidity worthy of a magical little being, the crystal grows, adding more than ten billion molecules a minute, poising them one upon another and creating a highly formalized construction in space. As a reverser of disorder the sugar crystal has gone several steps further than Maxwell's imaginary being; it has both selected and built a more complex whole.

This amazing process of crystallization is not a small or isolated phenomenon. It has formed much of the earth itself, the inner planets, and many objects farther out in space. The solid materials that make up the world around us are almost all crystalline in nature. Magnify any little pebble or grain of sand and you will see that it is a single crystal or a collage of tiny crystals. Each one is an exquisitely wrought design like a miniature jewel with shining facets. The entire lithosphere is woven of crystals—granite and limestone, copper and tin, diamonds and ice. Rich stores of treasure—gold, rubies, and emeralds—have been laid down by this process in the crust of the planet.

Most of the gold on earth is widely and diffusely distributed. It is present in all the rocks of the continental crust, but so thinly scattered that it represents only one part in 250 million. Although gold is slightly more abundant in the rocks of the ocean floor and in sea water, the concentration is still extremely small—a condition approaching maximum entropy. However, as some lucky prospectors have discovered, it has been concentrated by crystallization in certain favorable places beneath the planet's surface. Veins of gold lead in winding paths down deep in the roots of ancient volcanic mountains. These mother lodes of precious metal are formed when hot fluids containing dissolved minerals force their way up from the warm layer of earth beneath the lithosphere, seeping through channels left during the solidification of volcanic rocks. Narrow fissures provide the most advantageous conditions for crystallization because they offer a large surface area in relation to the volume of flow; in wide passageways the fluids pass through without depositing much ore. As these solutions, containing many different kinds of atoms, percolate up through slender openings toward the earth's surface, temperature and pressures drop. At the critical stage the presence of just a single crystal, a particle of other material, or perhaps a small protrusion on the rock wall can start the transformation of formlessness to form. Suddenly as though a magic wand had been waved, the atoms of gold leap from the amorphous liquid state and, joining together, construct that beautifully ordered work of art—a crystal of pure gold. Untold billions of these gleaming flakes fill the labyrinths of the rich veins that have been discovered around the world and bear shining testimony to the power of nature to select, to differentiate, to create order where there was none before.

Other natural processes also involve selection—we have noted a number of them in earlier chapters of this book. Some types of matter can act as a mold for other organisms. Clay, for example, serves as a template for protein molecules. This action depends upon its ability to select certain molecules from the wide assortment of those that impinge upon it and to hold these on its surface long enough for bonds to be formed between the chosen units.

All catalytic reactions involve a kind of natural selection. In such cases a *process* is selected. For example, the presence of nitrogenase enables certain bacteria to fix nitrogen without the infusion of the extravagant amounts of energy usually needed to make this essential element available in a form useful to living things. The catalyst is not used up in this reaction and—like Maxwell's Little Demon—it remains on the job indefinitely, directing the flow of chemical changes in a way that is economical of energy and advantageous to life. Nature makes use of catalytic processes in a great variety of ways. Enzymes (the catalysts of life) control differentiation and orchestrate the activities that maintain every organism.

But the most outstanding example of selectivity is the molecule which carries the genetic code—the molecule that makes reproduction possible, and vastly accelerates the extension of Form in space and time. A single strand of DNA contains molecules known as nucleotides (which are composed of sugar, phosphate, and organic bases) joined together like beads on a string. Each of these nucleotides selects from the assortment of molecules that are circulating within the cell the partner that best completes its own spatial configuration. This selection is not haphazard; it is perhaps the most precise of all natural processes. *A* always (or almost always) chooses *a, B* chooses *b* in exactly the proper sequence. When all the partners are lined up, each to its natural mate, they are linked together into a chain by the action of appropriate enzymes.

In this way strand *x* acts as a mold for a complementary strand *y*. Both *x* and *y* are capable of standing alone and serving as molds for new complementary strands—*y* molds another *x, x* another *y*. The single DNA molecule becomes two, then four, then eight, and so on in geometric progression. Each one of these molecules, containing at least ten billion atoms, is arranged in a precise pattern. Even more wonderful than the gleaming crystal of gold which is created by repeating a small regular configuration over and over again like a standard tile floor, the DNA molecule is constructed in a much more intricate and less repetitive fashion. It is composed like a mosiac, a single coherent masterpiece of design.

The DNA molecule is also characterized by a remarkable degree of permanence. These strands of genetic code are replicated millions of times, generation after generation and century after century, with only rare and relatively minor alterations. On the average perhaps once in a million replications a mutation occurs, and this is usually the result of impact from an external source. Even then the change is often reversed by a return mutation. Some species, like the horseshoe crab, have existed basically unchanged for hundreds of millions of years. As physicist Erwin

A single coherent masterpiece of design.

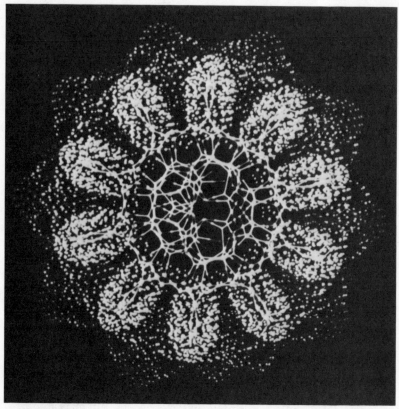

Plate 11
Computer Graphics – generated Axial View of a DNA Molecule (Courtesy
of Computer Graphics Laboratory, University of California).

Plate 12
Rose Window in Washington Cathedral Depicting the First Chapter of
Genesis (Photograph by John Grupenhoff, courtesy of *Science*).

Schrödinger said, the DNA molecule "displays a durability or permanence that borders on the miraculous."

This observation brings us to the third way in which organisms reverse the trend to disorder described by the law of entropy. It is an activity best described as *self-preservation*. Universally recognized in all living things, this trait is also exhibited on a more primitive level by inorganic matter. An atom or a molecule arranges its component parts to achieve a state of maximum stability in its environment, and when this arrangement has been disturbed by external forces it restores the original configuration as rapidly as possible.

Consider, for example, the atom that has been bombarded with energy. It enters an "excited state" with one or more of its electrons arranged in wider, less stable orbits. Very quickly this additional energy is rejected and the atom drops back to its most stable, ground state. According to the Second Law disorder has increased in this spontaneous change because the energy discarded by the atom has been turned into heat or radiation and is less available for doing work.

But something else has happened, too. As a result of returning to the lowest energy level the life expectancy of the atom has increased. In its lowest energy state a larger quantum (a packet of energy) is required to disrupt the organism. In an excited state, on the other hand, the atom already possesses some of the extra energy needed to effect this disruption; so it can be knocked out by the addition of a much smaller quantum.

The knockout blow that could destroy the organism comes from random encounters with other bits of matter or radiation in space. Each atom and molecule is immersed in a medium of randomly moving particles; the higher the temperature, the more rapid the motion of these particles. The survival of the atom or molecule depends on two factors: the quantum of energy needed to disrupt it—let us call this Q for brevity—and the chances of receiving this blow in the given environment. These chances increase with the temperature of the medium. By rearranging its component parts (usually with the rejection of some energy) the atom or molecule can effect large changes in Q while changes in the temperature of the medium caused by this action are relatively small. To give an example cited by Erwin Schrödinger, doubling Q can change the life expectancy of a molecule from one tenth of a second to 30,000 years although little change of temperature is involved. From these figures it can be seen that the drop to a lower internal energy level can greatly increase the stability of the organism. These spontaneous changes are occurring continually throughout all levels of complexity in the universe. The Second Law measures the increase of temperature caused by them but does not take into account the resulting increase in stability of orderly arrangements of matter. If life expectancy were included in the calculation it would be obvious that many spontaneous natural events cause an increase in order when measured in all four dimensions—time as well as space.

As we have seen in human populations throughout this last century, a reduction of the death rate while other factors remain constant results in an increased world population. Following this same principle, we can conclude that the universe is more densely populated with units of form because all organisms have the power of maintaining themselves and regenerating their form when it has been disturbed.

The creative force of self-preservation is a characteristic of organisms but not of arrangements imposed upon a system from outside. A carbon atom, a sugar molecule, an ice crystal exhibit this trait but a house, a steam engine, or a garden do not. What is this "undefined but precious something," as Donald Culross Peattie called it, "that makes the ugly thousand-legged creature flee from death?"

The precious power of self-preservation is born into each one of us. We know and understand it from direct experience. I was made poignantly aware of this truth many years ago when my youngest child was four years old. She was desperately ill with double pneumonia, a viral type that did not respond to antibiotics, and for eighteen hours her temperature hovered around 105°. Her little body looked blue but was radiating heat like an oven. I wrapped her in wet towels and sat beside her bed throughout the long night hours, listening to her labored breathing. Finally around dawn I got up to leave the room for a minute. But as I stood up, she caught hold of my skirt. "Don't go away, Mommy," she pleaded. "Stay here and keep me . . ." Her voice trailed off and I leaned down close to catch the words: "I . . . am afraid . . . I can't keep myself."

To keep oneself is so powerful a drive it overcomes almost insuperable obstacles. It calls up great courage and acts of fortitude that stir our pride in the human species. It enables the prisoners to return from death marches and concentration camps. It makes the mistreated slave cling to life in spite of pain and despair. It sustains the lone survivor of a plane wreck on the ice floes of the Arctic. And yet self-preservation is not a purely human trait; it permeates all the levels of form in nature. The delicate molecules of organic matter tumbling in the dark reaches of outer space resist dissolution, reject the extra energy forced upon them by chance encounters, and return to their most stable state. The captured fly struggles to release itself from the spider's web. The newborn sea turtle makes a dash for the relative safety of the open sea. Wherever there is a self-organized unit of matter, self-preservation is a fact to be reckoned with. At the lowest levels only a limited number of responses are possible, but as complexity increases more options emerge. With living beings the ability to reproduce adds an important new way to preserve one's form. By maintaining the continuity of the species the existence of the basic pattern is greatly extended in the time dimension.

We have observed how the single slime fungus cell, threatened by scarcity of food, joins with others of its kind in building a fruiting body, producing spores capable of growing into new individuals. Although many of the old cells die in this process, the life of the species is preserved.

Spontaneous rearrangement of internal parts to create a reproductive organism is common in nature. One of the most fascinating ones is the metamorphosis of a caterpillar into a butterfly. Joseph Wood Krutch described the process in these words:

> Once the skin [of the caterpillar] has been shed and the
> creature—or what is still left of it—ceases to move, the

destruction of the original organs and the fashioning of new ones goes on apace. Free-moving cells, much like the phagocytes or white blood corpuscles in the human body, absorb and carry away the disintegrating material, and at the same time new organs begin to form. Ever since the day when the caterpillar was hatched from the egg it has carried within its body certain little groups of cells which were useless until now. They are the buds, if the term be permitted, from which the butterfly's organs will develop, and these organs grow on the material the phagocytes have been carrying away from the parts of its dead self. From these buds a butterfly begins to form as the caterpillar was dissolved. No new material save perhaps air and moisture is available any more than anything is available to an egg closed within its shell. What is more remarkable, is that almost nothing is left over. The material in one caterpillar is just sufficient to make one butterfly!

"Almost nothing"—the qualification is interesting. If you have kept your butterfly under continuous observation from the time it ruptured the skin of the chrysalis until the moment when it took wing you probably observed, at some moment not long before the last, that one or two drops of liquid fell from its rear to the ground. "What," you may have asked, "is that?" It was Nature's miscalculation; or rather, her margin of safety. Being sure to have enough—and she cannot predict just how much was going to be lost by evaporation from the chrysalis— she had one or two drops left over.

The fact that a few drops have been left over and rejected as waste might lead to the conclusion that the process of metamorphosis as a whole has resulted in a decrease of order in this biological system. But like calculations based on the Second Law, this interpretation would not take into account the higher level of form which emerges from the process. The butterfly is capable of reproduction while the caterpillar is not. This act enables the organism to multiply and extend its generic form in space and time.

A small amount of material is left over in any of the natural transformation processes. But this material is not wasted, as we have seen. Even the random energy of heat motion is useful, providing the warm gentle environment in which living organisms can flourish. Although random thermal motion has become more evenly distributed as the universe has evolved, and average temperatures have steadily declined, these changes have set up conditions which actually aid the creation and preservation of Form. Until the temperature had fallen from the original peak at the first moment of creation, even the simplest fundamental particles could not take shape; until it had fallen further neutral atoms could not form and maintain their existence in space. Molecules immersed in a high-temperature medium have a lower life expectancy than the same molecules in a cooler environment, and the hot fluids carrying minerals in solution must cool to a certain critical level before elements like gold or silver or copper crystallize out. Although high temperature differences are efficient in running mechanical devices, they do not set up the most favorable conditions in which Form can grow and evolve.

Leftover matter and energy are not eliminated from the form-building process. The waste material from one type of organism serves as nutrient for another—as fish pellets nourish the zooxanthellae on the coral reef. We know that recycling like this takes place throughout the biosphere and probably also throughout the cosmos. Nature maintains a remarkable balance in all these exchanges and reuses everything. Recycling is an important part of the evolution of Form; as long as creation is not complete, it provides the raw material needed to build the next phase and the next.

Death and dissolution are inevitable aspects of this process. If death did not occur the substance of life would have been frozen in the earliest simple stages of formation. The infinite extension of each individual in the time dimension would bring to an end all evolutionary change. Our instinctive resistance to dissolution is a manifestation of the basic drive for self-preservation, but even the most confirmed egotist would not desire infinite extension in the spatial dimensions. An individual life has form in all dimensions. It has depth and breadth; it soars at times to great heights, and when death comes a well-filled life is rounded out in time—completed, not finished, because nothing that has been can be nullified. It remains an essential part of the tapestry that is being woven in the warp and woof of space and time.

We have seen how eon after eon more elaborately integrated, more finely differentiated units of matter and energy have been born out of the matrix of simpler things. Quarks sprang from the undifferentiated plasma of the infant universe; organic molecules were synthesized in the dark reaches of outer space; and in the ancient seas tiny one-celled organisms joined together to construct the fantastic underwater castles of the coral reefs, where blue angelfish and golden rock beauties find protective shelter and purple seafans lift their graceful bodies high enough to catch the sunshine.

Adherents of the Second Law (in its application to the cosmos) maintain that the examples of order we have found in nature are only small local phenomena; in other places disorder has increased. To prove there has been a net increase, they say, you must consider a totally closed system. But completely isolated systems are rare indeed. In nature everything is connected to everything else. The universe itself is the only really isolated system. We have traced its evolution from a mass of disorganized plasma through a long series of increasing levels of complexity, both on the grand cosmic scale and on the microscopically small— to galaxies and planets condensed from stardust—to the DNA molecule and the mind of man. This is not the life story of a universe running downhill toward a completely uniform and random state of maximum entropy in which no events can occur.

All of these observations lead to the conclusion that the Second Law of Thermodynamics, useful as it is in the whole range of phenomena for which it was originally conceived, cannot be used to evaluate the evolution of Form throughout time. The law of increasing entropy allows us to quantify all physical processes depending on the irregular motion of molecules and atoms. It predicts the inevitable dissolution of the orderly arrangements that are imposed upon matter—a pyramid, a mosaic, a monument carved in stone. But it does not give us useful information about the future of a tulip bulb, a butterfly, or the universe itself.

I am aware that in making this argument I am challenging the interpretation of one of the most important laws of physical science. It would be safer to avoid trespassing on this controversial ground. But I believe that the issue must be raised, because the dogma of a universe running downhill is blocking the development of more positive views of the physical world. Many writers who in recent years have recognized the importance of organic wholeness in nature (Edmund Sinnott, Ludwig von Bertalanffy, Ralph Lillie) have restricted their theories to the world of living things. In this position they did not need to challenge the postulate of a cosmic trend to disorder. Although life produces orderly events, the phenomenon is small enough to be concealed under the vast outflow of "positive entropy" into space. These scientists might have broadened their theories to encompass the whole physical world if they had not been influenced by the interpretation of the Second Law which concludes that the cosmos is becoming more disorderly with the elapse of time.

Teilhard de Chardin's great theory of the creative forces working throughout time was also circumscribed by the need to stay within limits imposed by this law. Since every synthesis costs something in terms of energy, Teilhard said, the "material concrete universe seems to be unable to continue on its way indefinitely . . . but traces out irreversibly a curve of obviously limited development."

Such a limitation undermines Teilhard's grand scheme of a universe moving toward higher degrees of organization and increased variety. If every synthesis reduces the totality, the whole is constantly diminished as it evolves.

If, on the other hand, we take into account the many wonderful ways in which energy left over from every spontaneous change can be swept up again into form-building processes, we do not have to accept this diminished conceptual scheme. The stuff of the universe is continuously recycled as the creative process is worked out. Like a butterfly taking shape within its cocoon, the parts are rearranged, the material reused to construct a more finely designed and more beautiful whole. In order to make sure of having enough, there may be a small amount of something left over in the final synthesis. In every creative process the exact amount of material needed for the finished work of art cannot be foreseen. But the remarkable economy of nature suggests that the totality will include most of the substance of the universe when the final Form emerges from the chrysalis of time.

CHAPTER 7

C HANCE AND NATURAL SELECTION

Men argue learnedly over whether life is chemi-
cal chance or antichance, but they seem to forget
that the life in chemicals may be the greatest
chance of all, the most mysterious and unexplain-
able property of matter.

LOREN EISELEY
The Firmament of Time

One school of modern artists has been experimenting with a new technique of painting a picture. Standing several feet back from a canvas they pelt it with pigment, so the little drops fall in a haphazard arrangement of dots, dribbles, and splashes on the flat surface. We can imagine that the overwhelming majority of these compositions turn out to be failures as works of art and the conscientious artist quickly disposes of them. But once in a long while the paint lands in a way that creates a pleasing pattern of shape and color. This one is selected and preserved. The chances of a fortuitous result increase with the number of tries; given enough tries, in fact, every possible combination of color and pattern would be produced. If the experiment were continued over untold generations, perhaps the trillionth time would re-create Leonardo da Vinci's *Mona Lisa* or Van Gogh's *Self Portrait!* In this way time can bring about arrangements of great improbability, but unless there is some mechanism for recognizing and selecting out the rare favorable ones, the entire process will produce nothing of lasting value.

The twentieth century scientific world-view is similar to this accidental artistry. Any form or order that we find in the universe is described as only a chance accumulation of substance—nothing more than a few bubbles of matter that have assumed a fortuitous shape, just one of a hundred billion possible arrangements thrown up in the turbulent sea of cosmic change. Add to this thought the concept (derived from the Second Law of Thermodynamics) of a blindly running flux of disintegrating energy and we complete the nihilistic view of the universe that has set the tone of modern thought.

The philosophy of a universe based on chance was born in the early years of our century when the German physicist Werner Heisenberg discovered that it is impossible to measure simultaneously the position and velocity of certain very small particles. This measurement is impossible in theory, not simply a matter of technique. For example, the very act of observing an electron in order to measure its position and velocity interferes with it sufficiently to cause errors of measurement. This discovery shook the very foundations of physics, which—up to that time—had been proceeding confidently on the assumption that precision of measurement was limited only by the instruments and methods available. By improving these the degree of accuracy could be increased without limit. Further refinement could always add another decimal place.

Like the Second Law of Thermodynamics, the uncertainty principle was extended to cover a wider range of phenomena than the ones for which it had been conceived. The fact that the behavior of fundamental particles cannot be exactly described or predicted was interpreted as evidence that the world is not ruled by strict causality. If a sequence of cause and effect cannot be directly observed then perhaps there is none. Electrons may enjoy a kind of free will. This concept, rechristened the "principle of indeterminacy," was welcomed by cosmologists and theologians who had been unable to accept the old mechanistic world-view where every event was seen as the inevitable consequence of the events that went before—every action was predestined and no free will could exist.

But while the principle of indeterminacy seemed to offer freedom, it landed natural philosophy on the other horn of the dilemma. If no true cause and effect relationship can be established, how can we account for the regularities—the laws—that science has been uncovering for many years? The answer seemed to lie in what has been called the laws of chance.

Physicists and chemists deal routinely with very large numbers of small particles engaged in disorganized thermal motion. In a liquid solution or a gas the path of each individual particle cannot be predicted, but because great aggregates of particles are involved, the random factors average out and a certain regularity is revealed. A chemist can predict that one minute after a particular reaction has started half of the molecules will have reacted, and after another minute three-quarters of them will have done so. When the process is studied in this manner disorder appears to produce order, and a statistical law of chemical reaction emerges.

In a similar way it is possible to predict accurately the pressure caused by a volume of gas at a known temperature even though the heat energy is distributed in a random manner among the molecules. From these observations it was only one small step to the assumption that all the regularities found in the physical world can be interpreted as the result of vast numbers of random events. In fact it might be said that law arises only statistically, due to the cooperation of myriad chances at play. Surprising as it may seem, this conclusion has been espoused by many physical scientists.

But if we examine statistical methods in detail, we find that the regularity that emerges from them is a measure of the element or elements that are not random. As illustration take the very simple case of tossing a coin. When it is tossed many times heads turns up as often as tails. The chance of getting heads is one in two because there are two sides to the coin. If there were six sides, as in a die, the chance of getting one side would be one in six. The factor that is not random in these cases is the shape of the coin or die. If the shape were also random, changing at each throw to three sides, four sides, eight sides, and so on, there would be no constant factor and no "law" would emerge.

If we played such a game of chance blindfolded and found that after 500 throws six different results occurred with equal frequency, we would know that the object being tossed was a cube. Following the same line of reasoning, whenever the shuffling of random elements reveals a distinct trend in any one direction, we may conclude that there is a constant tendency at work. A broad look at the history of the universe and the course of evolution has shown that, in spite of the continual mixing of random elements, there has been a trend toward greater complexity and higher stages of organization of matter. The significance of this trend can be seen very dramatically in the evolution of living things.

According to the most widely accepted opinion, however, the process of natural selection is interpreted as the inevitable consequence of the chance elements produced by mutation. This point of view was presented by Julian Huxley in his classic book *Evolution in Action:*

> How can a blind and automatic shifting process like selection, operating on a blind and undirected process like mutation, produce organs like the eye or the brain, with their almost incredible complexity and delicacy of adjustment? How can chance produce elaborate design? In a word, are you not asking us to believe too much? The answer is no: all this is not too much to believe, once one has grasped the way the process operates. Professor R. A. Fisher once summed the matter up in a pithy phrase— Natural selection is a mechanism for generating an exceedingly high degree of improbability. . . . The clue to the paradox is time. The longer selection operates, the more improbable (in this sense) are its results; and in

point of fact, it has been operating for a very long time indeed.

A little calculation demonstrates how incredibly improbable the results of natural selection can be when enough time is available. Following Professor Muller, we can ask what would have been the odds against a higher animal, such as a horse, being produced by chance alone; that is to say, by the accidental accumulation of the necessary favorable mutations *without* the intervention of selection. To calculate these odds we need to estimate two quantities—the proportion of favorable mutations to useless or harmful ones and the total number of mutational steps, or successive favorable mutations, needed for the production of a horse from some simple microscopic ancestor. A proportion of favorable mutations of one in a thousand does not sound much but is probably generous . . . and a total of a million mutational steps sounds a great deal, but is probably an underestimate. . . . However, let us take these figures as being reasonable estimates. With this proportion, but without any selection, we should clearly have to breed a thousand strains to get one containing two favorable mutations; and so on, up to a thousand to the millionth power to get one containing a million.

Of course, this could not really happen, but it is a useful way of visualizing the fantastic odds against getting a number of favorable mutations in one strain through pure chance alone. A thousand to the millionth power, when written out, becomes the figure 1 with three million naughts after it, and that would take three large volumes of about five hundred pages each, just to print! . . . No one would bet on anything so improbable happening; and yet it *has* happened. It has happened, thanks to the workings of natural selection and the properties of living substance which make natural selection inevitable.

What are the properties of living matter which make natural selection inevitable? They are the characteristics of self-preservation, regeneration, and reproduction. These properties work to stabilize and sift the new forms created by mutation. Since favorable ones are those that can be stabilized successfully, these are automatically selected and act as a base for the next mutational novelty. The selection and retention of the earlier steps changes dramatically the chances of accumulating the necessary favorable mutations.

The role of selection in improving probabilities can be illustrated by an analogy from card playing. In a poker game a player is dealt five cards and is allowed to discard any number of these and redraw. If in his original hand he holds three cards of the same suit, say spades, he might decide to try for a flush. Discarding the other two cards and drawing once, his chances of completing the flush are about one in sixteen. If he were allowed to discard all the unwanted cards and redraw many times, it is obvious that the larger the number of draws (N), the greater the probability of achieving a flush. As N increases, the chances become so large as to be inevitable. Under these circumstances more time allows more and more improbable combinations to occur. But note that this is true only because each time the player is allowed to select and retain the spades that have come into his hand. If at each draw he were required to throw back the whole hand, then on the Nth draw he would have no better chance of holding a flush than on the first. Time in this case would not create states of greater improbability.

There is an interesting law of chance which says that if you want a number of lucky events to occur all at once or in immediate succession, you must multiply together the chances of each one happening separately. The probability of drawing one spade from a full pack is one in four, because there are four suits. But if you would like to draw five spades in succession (let us assume a full deck in each case for the sake of simplicity), then you must multiply one in four by itself four times. You will find that the odds against you are 1024 to one—a much poorer gamble than the poker hand where choosing, holding back, and redrawing produced favorable odds. Remember, too, that in this case you have asked for only five favorable events, while in the mutations that produce living organisms millions of events are involved. It is apparent that chance could not have produced the remarkable associations of order that we find in nature if there were no universal tendency to select and preserve the fortuitous combinations.

Natural processes work in a selective manner, stabilizing and extending the orderly arrangements which have come together by random action. The very origin and evolution of life are dependent upon these characteristics of matter. As a result of mixing over long periods of time, the proper elements to form a living cell would have encountered each other by chance; but if these elements had not combined into a single self-preserving unit of matter, they would have drifted apart again and the combination would have been lost. In fact, this event itself could never have occurred without the chemical evolution of stable configurations of matter up from the single electron to the complex proteins and amino acids. Each more elaborate combination is statistically more improbable; it only becomes possible because the steps before it have been preserved. The mysterious quality of integration present in all organisms is a kind of Maxwell's Little Demon, retaining the units of form, building throughout time increasing degrees of complexity, and reversing the trend toward states of greater randomness predicted by the Second Law of Thermodynamics.

While acknowledging that natural selection is a factor in the realm of living things, many biologists and naturalists are uncomfortable with the theory that evolution is caused only by accidental external forces. They know that it is influenced by factors within the organisms themselves. "The tale as told by the careful expositors," declared Joseph Wood Krutch, "is a very tall one and there are few who will not admit that it would be easier to swallow if only we could believe that it was not exclusively a story of miracles which pure chance, complete blindness, and utter mindlessness are supposed to have accomplished."

Natural selection is not the product of chance alone. It is the result of form-building properties in nature integrating and stabilizing the favorable changes that are produced by chance. The tendency of each organism to extend its existence in time—its ability to arrange its component parts to maximize its life expectancy—results in the retention of successive layers of more and more complex order, so each step can be built upon the preceding ones. Competition becomes an important factor in cosmic evolution at the point where organisms are capable of reproduction. Life forms replicate themselves abundantly and press against a limited food supply. The fittest compete most successfully for the resources; they live longer (on the average) and are able to reproduce themselves more effectively than the less fit. But the traits of self-preservation and regeneration existed before the first living cell was formed, and long before competition for food and living space became a selective force. The atom recaptures its lost electrons, the crystal restores its fractured shape, the molecule discards the disturbing energy forced upon it by random encounters. These traits become increasingly important as new abilities are generated by greater complexity. The spectacular success of living matter causes competitive pressures which automatically select those forms that can best extend themselves in space and time.

The process of mutation which provides the mechanism for evolutionary change does appear to be largely a random phenomenon, although even here properties of self-organization influence the range of variations that can be produced. Let us listen to the German geneticist Ludwig von Bertalanffy on this subject:

> Since the gene is a physico-chemical structural unit of the nature of a large protein molecule, and since a mutation represents a transition to a new stable state . . . a change will certainly be possible in a number of directions but not in all directions, in a similar way as only certain quantum states are permitted to an atom. In both cases, quantization is at the basis of the jump-like character of the change, as well as of the high stability and organization of the system. An atom cannot take up any small quantities of energy it is exposed to under the constant bombardment of the heat movements of the surrounding particles; only a complete quantum jump will induce a change in it. This ensures that it may remain unchanged over an indefinite time. In the same way, the "quantized" character of mutations is at the basis, first, of their dis-

continuity, secondly, of the great stability of the gene and the relative rarity of mutations, and thirdly, it follows that the number of possible mutations is not infinite, since only certain stable states are "allowed."

Here we come to an important problem. . . . We find no evidence either in the living world of today or of past geological epochs for a continuous transition (from simpler to more highly organized forms). What we actually find are separate and well-distinguished species. Even the existence of more or less numerous mutations, races, subspecies, etc., within the species does not alter the basic fact that intermediate stages from one species to another which should be found if there were a gradual transition, are not met with. The worlds of organisms, living and extinct, do not represent a continuum but a discontinuum.

The discontinuity of species is based presumably on the fact that certain conditions of stability exist not only for the individual genes but also for genomes [an interacting system of genes]. . . . A "species" represents a state in which a harmoniously stabilized "genic balance" has been established, that is, a state in which the genes are internally so adapted to each other that an undisturbed and harmonious course of development is guaranteed. If there are no external disturbances, stability is ensured for a theoretically unlimited number of generations. If a mutation occurs it means a disturbance of this pattern; therefore in the majority of cases, mutations will lead to unfavorable, even lethal, results, even before selection is taken into account. But every gene acts not only in its own right but more or less also as a modifying factor influencing the action of the rest of the genome. . . .

Thus the changes undergone by organisms in the course of evolution do not appear to be completely fortuitous and accidental; rather they are restricted, first by the variations possible in the genes, secondly, by those possible in development, that is, in the action of the genic system, thirdly, by general laws of organization.

Like a flight of stairs, the process of evolution moves toward ever higher levels, the changes occurring in steps because only certain states are harmoniously stabilized. Furthermore—and this is a factor that was not fully appreciated in the early 1950s when von Bertalanffy wrote—large advances have been achieved when the synthesis of separate units into more complex systems has taken place. For example, when single-celled organisms united to make a eucaryotic cell, the change represented a quantum jump.

The integrating qualities displayed by the gene and the genome are characteristics of wholeness that we have observed in all organisms, both living and nonliving, but they are most apparent at the stage of complexity represented by living things. It is not surprising, therefore, that the concept of wholeness has been pioneered by scientists involved in the study of life, while the physical sciences have achieved their great successes by dissecting and isolating the parts of any system for detailed examination. The analyst has a great advantage over the synthesizer, for it is easier to take something apart than to put it back together again. But to understand a whole we must know more than what pieces constitute it; the dynamic integration of these components into a single unit must be understood. Like Humpty Dumpty in the nursery rhyme, the whole cannot be restored once its integrity has been destroyed.

At the highest levels of complexity the properties of wholeness are recognized as manifestations of the unity and integration of the individual. It is obvious that they are not caused by mechanical forces. No one would attempt to measure the "force" that causes self-preservation in human beings. The cause is more accurately described as a nisus—a striving of the individual to preserve itself in the face of a changing environment. In a similar way the adaptation of every organism involves its active participation. It reorganizes its inner structure and assumes the most favorable position in relation to its environment. Various degrees of this ability are manifested throughout the entire range of organisms.

Each organism has a certain latitude of freedom. The simplest ones can act in only a limited number of ways, but as complexity grows, the options available to the organism increase. The degrees of freedom are multiplied as we move up from atom to molecule to gene to living organism. Living things exhibit an ability to adapt themselves to a given environment, to seek out more favorable environments, and even to alter the environment to suit their needs. In mankind the degree of freedom is very large, large enough to change—even to jeopardize—the whole earth system.

The ability of each organism to respond to environmental challenges and to new opportunities introduces a degree of uncertainty into the physical world. It is not possible to predict the response of each individual organism, even at the very elementary level, but where large numbers are involved the actions average out and the behavior of the whole assembly can be predicted statistically with some degree of reliability, just as the laws of physics state.

The uncertainty, however, does not result from meaningless, random action. Certain principles are followed throughout leading to increased stability, increased complexity, and the multiplication of units of form. Although a vast experimentation is taking place which involves degrees of freedom at many different levels, the direction of the process is constant and predictable.

"Free Will and Determinism are simply misunderstandings of history," Jacob Bronowski said. "History is neither determined nor random. At any moment, it moves forward into an area whose general shape is known but whose boundaries are uncertain in a calculable way. A society moves under material pressure; but at any instant, any individual may, like an atom of the gas, be moving across or against the stream. The will on the one hand and the compulsion on the other exist and play within these boundaries."

The recognition that some measure of free will, some ability to act, exists far down the scale of self-organized units of form suggests an answer to another question that has concerned philosophers for many years. Consciousness is the central experience of life, and many people have assumed that it is unique to human beings. But animals appear to share this trait to some degree. A dog knows its name and responds when it is called. Most significantly, the dog acts to protect itself; so do a centipede and an anemone. Even the elementary inert forms of matter act in a manner which extends their own existence in time. Surely self-preservation would not be possible without *a sense of self*. Perhaps consciousness, like integration and the ability to act, is present (in a very rudimentary sense) even in the most fundamental organisms. This trait expands and reaches more sophisticated stages of expression as higher levels of complexity are reached. When units synthesize to create a larger organism, the sense of self takes another step up the scale. As Form continuously increases, encompassing larger and larger segments of the cosmos, the self becomes greater, the many become one. And through this creative process the universe is becoming conscious of itself.

CHAPTER 8

MAN'S PLACE IN THE PROCESS

*Man is a rope stretched between the animal and
the superman—a rope over an abyss.*

FRIEDERICH WILHELM NIETZSCHE
Thus Spake Zarathustra

There is no lovelier image in our Western heritage than the
legend of the first man and the first woman wandering naked
and carefree through the Garden of Eden. The newly created
world was virginal and pure and fresh as an early spring day
when the trees are in flower, the leaves still tightly curled and
no larger than rosebuds. Violets and periwinkle and morning
dew bejewel the grass that is verdant with an intensity like the
green flash of light from the setting sun as it slips down behind
the gently curving edge of the western sky.

Death had not yet cast its shadow across the scene—it had not
blackened a single leaf nor sullied the sweet fragrances of emerging life. The story of Adam and Eve is a beautiful allegory sym-
bolizing the state of innocence from which we came, but to

which we can never return, because we are free in a way that no other organism has ever been—free to choose, to pick the apple from the forbidden tree and to become as gods knowing good and evil. Without freedom there is no good or evil.

In mankind the long process of evolution which had been moving persistently in the direction of increasing awareness, of more power to master and mold the environment, has produced a creature capable of going against the stream or moving joyfully with it—free to accelerate vastly the form-building process but also free to obstruct it. The independence that caused mankind's expulsion from the garden of innocence carries with it a heavy burden of responsibility. In the words of Feodor Dostoevsky, "Nothing has ever been more insupportable for a man and a human society than freedom." Its price is an inner world of fear and insecurity and loneliness. And yet this power to choose has given mankind his unique place as the spearhead of the creative process, making him the most active participant and a partner as well.

How and when did the transformation occur that turned a primitive primate into the thinking, striving, brawling, loving, and dissenting creature that is man? Did it take place by slow accumulation of many small mutations (as Julian Huxley postulated for the horse), each one chosen out of many possible variants by natural selection? Or did it occur in a decisive quantum leap? The information available from studies of blood chemistry and the genes as well as from a detailed examination of the fossil record suggests that the evolution of man involved both of these phenomena. First a decisive change marked the divergence of *Homo* from the great apes; then a long period of gradual change occurred in which survival of the fittest selected the most favorable individual variations as the potential inherent in the new genome was explored and expressed.

In an earlier chapter we examined the latest scientific evidence which showed that the changes in genetic code from chimpanzee to man were remarkably few in number. Chromosome No. 2 in human beings seems to have been formed by the fusion of two smaller chromosomes carried by the gorilla and the orangutan as well as by the chimp. Two other chromosomes, Nos. 1 and 18, differ from those of the chimp only by small inversions—a rearrangement of materials already present. It is tempting to compare these changes with the synthesis and the active adjustment of parts that takes place in the formation of many simpler organisms. According to traditional theory, the chromosome changes must be explained as the result of impacts of random energy, the most favorable being selected from many unfavorable ones. But in this case there is no evidence that many other variants were produced or that the competition for natural resources at that time was very great. Probably both chance and self-integration were involved—a random impact of energy caused an alteration and then the whole genome adjusted its component parts to achieve a stable state of genetic concordance. At the present state of our knowledge it is not possible to answer definitively the question of how the modifications in the genetic code occurred, but there is general agreement on the timing. The hominid strain (the line leading to man) seems to have diverged from its closest relative, the chimpanzee, between four and ten million years ago.

This estimate is further supported by studies in molecular biology. Differences in the structure of basic body proteins in two different species, for example, man and gorilla, can be counted and the number of mutations necessary to produce such alterations can be estimated, as well as the average length of time for each mutation to occur. Comparisons of this kind have shown a remarkable degree of similarity between human and gorilla hemoglobin and certain proteins of the immune response system. There is an even closer, almost perfect match in man and chimpanzee. Based on these results the time of separation between chimp and *Homo* may have been as recent as four to five million years ago.

Since Darwin's time and until very recently, evolutionists have assumed that the divergence of hominid from primate stock must have coincided with the appearance of the large brain which is the key feature distinguishing *Homo sapiens* from the rest of the animal world. But anthropologists examining recently discovered fossil skulls have reached the surprising conclusion that the brain increased very slowly in size until about one million years ago, and then its evolution became explosively rapid. For several million years the creature that carried within its cells the seeds of abstract thought, of dreams and poems, of knowledge and frightening power, wandered innocently through the lush forests and across deeply vegetated savannahs, plucking fruit from the trees, capturing small animals to eat. His physical appearance underwent gradual alteration—the legs became longer, the heavy jaws were lightened and refined, the pelvis and feet were modified to accommodate upright posture. The brain case also grew, but by one million years ago it was still only half the size of modern man's. While the hominid line was evolving the great apes remained basically the same.

We can imagine that in the early years primate populations were relatively small and resources were abundant, so competitive pressures were slight. However, those individuals who were endowed by chance with the best brains were able to invent and develop certain skills that gave them a position of dominance and more opportunity to pass on their genes to the next generation.

The making and using of tools, for example, was one of the important discoveries. It is not, as once believed, a uniquely hominid accomplishment; a number of animals use simple tools. In the Galápagos Islands a finch has been observed extracting insects from holes with a cactus spine. In Africa Jane Goodall watched chimpanzees use twigs and blades of grass to fish termites out of their nests. She also observed male chimps carrying branches and brandishing them to frighten off their foes. Man and his immediate ancestors, however, were the only species that fashioned tools out of stone. The fossil record shows that as early as two million years ago apemen were chipping quartz and lava to make crude cobbles for chopping and bashing and sharp points for cutting and piercing. The ability to shape a blade and use it to skin an animal, for example, required manual dexterity and brains enough to recognize the advantage that could be gained by this innovation. Gradually the advantage led to natural selection of dexterity and intelligence. Fossils of apemen about two million years old show some increase in the size of the brain case, which is approximately twice that of the ancestral types. For the next million years, however, very little change seems to have taken place. We may infer that selection pressures were not great—that these early hominids were comfortably adapted to their environment, and there was no strong need to improve their abilities.

Then about one million years ago the evolutionary pace accelerated. Competitive pressures may have intensified, due perhaps to increased population or to a less favorable environment. One possibility that has been suggested is the deterioration in climate that occurred when the Ice Ages descended on the planet early in the Pleistocene Epoch (between one and two million years ago). By this time our ancestors had spread out to occupy large areas of the earth from South Africa to Europe and across Asia to China and Java. In many of these regions rigorous environments replaced the warm equitable habitats where food had always been plentiful. Competition for diminishing resources and more severe winters caused greatly increased selection pressures, favoring the development of intelligence. In order to survive early man learned to plan ahead, to lay in food supplies, to join in cooperative action with others of his kind, to find shelter in caves, and to cover his body with animal skins. And he also discovered the use of fire. The remains of hearths 500,000 years old have been uncovered in China.

The slow process of chance mutation and survival of the fittest might eventually have produced a primate with a heavy fur coat capable of surviving the severe winters. But by the use of intelligence—by the discovery of fire, clothing, and shelter—the problem was solved much more efficiently. As the advantages accruing from the use of intelligence became apparent, selection pressures for development of brain accelerated. Whether the key external impetus came from deteriorating climate (a number of biologists question this) or some other threat to survival, there was also an internal factor that must not be underestimated. By calling up the inner resources with which he was so richly endowed, early man built a new micro-environment in which intelligence became an ever more advantageous possession.

As the brain developed, the head increased in size to accommodate the larger number of brain cells. This change began to pose a serious threat to survival, for the infant's head could not pass easily through the birth canal, and difficult delivery threatened the life of both mother and child. But sometime in evolutionary history nature provided an unusual solution to this problem. The human baby is born in a relatively immature state. Almost all of the increase in human brain size takes place after birth; it trebles during the first year of life.

Although a greater degree of dependency at birth entailed some biological disadvantages, the positive gain of less hazardous delivery was important enough to tip the selection pressure in favor of this adaptation. Once it became established the brain could increase and develop rapidly.

Two other characteristics that appeared about the same time were also related to the immature state of the hominid baby at birth. The foetus of the chimpanzee in the late stage of pregnancy is relatively free of hair, just like human beings. This prenatal nakedness became an adult trait of mankind, even though it was a disadvantage to the female of the species. Babies of the great apes cling to their mothers' hair, and can be carried easily wherever the mother goes. Without body hair, the human female must carry her infant in her arms.

An extended childhood also became characteristic of the hominid line. The period of immaturity for the chimpanzee is about eight years but for a human child today it is almost twice that long. At first this change, too, must have been a biological liability, exposing the young to dangers from predators for a more protracted period of time. It is logical to conclude that these attributes went hand in hand with some overriding competitive advantage—like the development of brain.

There were prices to be paid for the wonderful package of power that brought with it freedom and knowledge. The adaptation of earlier delivery was a marginal one at best—as all women who have borne children can attest. Up until just a century ago childbirth was a leading cause of death in women. Even with all the aids of modern medicine it is a difficult and painful experience. "In sorrow," as the Bible says, "thou shalt bring forth children."

But long before the rapid enlargement of the brain had occurred, at least a million years earlier than the first evidence of
stone tool-making, our ancestors were beginning to construct cohesive social organizations. Males and females, old and young,
lived together, cooperating in the essential life activities for the
mutual benefit of all the members. It was within the context of
this social system that the most important advances toward hu-
manness were made.

In the Afar depression of eastern Ethiopia bones of at least
thirteen individuals including infants, juveniles, and adults have
been found at one site dated between three and four million
years ago. Very primitive hominid groups like this are believed
to have practiced division of labor and made special provisions
for the young—characteristics that are also evident in the social
organizations of other vertebrate species.

One of the interesting examples of community life is that of
the wild dogs that inhabit the same savannah regions in Africa
where our distant ancestors evolved. Although these animals are
fierce hunters and savage destroyers of their prey, they lead a
gentle and mutually protective group existence. Litters of newborn pups are sheltered in a den or burrow and are kept there
until old enough to fend for themselves. The young are fed and
cared for, not just by their own mothers but by all the members
of the community. One female guards the pups while the other
adults go off to hunt in packs. Their well-coordinated group attacks make the wild dogs the most successful and most feared of
all carnivores. But the food from a kill is shared with a remarkable degree of equality. Both males and females hastily ingest
large pieces of meat at the site of the kill and then return to the
den where they regurgitate a portion for the pups and the adult
that had stayed behind to watch the young. This behavior en-
sures that each member of the community, no matter how weak
or small, receives the nourishment necessary for normal development and activity.

All monkeys and apes also live in tightly knit social groups
containing from ten to two hundred members. Within these extended primate families affectionate regard for each other is a

marked characteristic. The mother and child bond is deep and enduring, and friendship between siblings lasts in some cases throughout life. Since multiple mating occurs during periods of estrus, males do not establish long-term relationships with individual females, and they do not recognize their own offspring. However, they do assume a generally protective role, guarding all the females and the young. When troops of gorillas and baboons move through the forest one or more mature males lead the group, and others take up the rear. If danger threatens, the males move to the attack while the females and babies take cover, climbing a nearby tree if one is available. The protection provided by the troop is absolutely essential for survival; an individual that drops behind or becomes separated from the group is spotted by predators and picked off.

These primate troops are nomads. Constantly on the move through the forest, they do not establish even a temporary home base like the den of the wild dogs. In the canine society there is more division of labor, since some adults must stay home to guard the young while the others hunt, and the food supply is shared more generously. The great apes very seldom share food with each other. In most cases even the mothers do not provide for their young after weaning. Only the chimps, our closest relatives, have been observed to occasionally divide a choice morsel with an infant or help him pick a piece of fruit.

In spite of these small differences, the almost universal need of living things to join with others of their kind and create a social structure is a striking phenomenon that can be explained very simply as a manifestation of nature's tendency to build higher and more complex levels of organization. A society is a larger organism in the process of becoming. It is an integrated system created by synthesis, and all of its components must be finely adjusted to work together.

In *Patterns of Culture* Ruth Benedict drew attention to the importance of studying cultures as articulated wholes, consistent patterns of thought and action. "Such patterning of culture cannot be ignored," she said, "as if it were an unimportant detail. The whole, as modern science is insisting in many fields, is not

Manifestations of nature's tendency to build ever higher and more complex levels of form.

Plate 13
The Maypole: Empire State Building, 1932. (Photograph by Edward Steichen).

Plate 14
Tourmaline Crystal (Photograph by Lee Boltin).

merely the sum of all its parts, but the result of a unique arrangement and interrelation of the parts that has brought about
a new entity."

The same theme was developed by Theodosius Dobzhansky,
who emphasized the dangers of interfering with the integration of
the social unit:

> Comparative studies of many kinds of societies, both non-
> literate and advanced, have disclosed that the beliefs, cus-
> toms, and practices of a group of people usually hold to-
> gether as related parts of a whole. A durable culture
> shows a certain patterning or integration of its compo-
> nents. A change in one part of a culture may, then, re-
> quire correlated changes in the whole pattern to make the
> system workable. The importance of this integration has
> been shown in a series of unpremeditated experiments.
> The contacts of primitive societies with Western civiliza-
> tion have in many instances resulted in a breakdown of
> the former and even in the extinction of the tribes which
> comprised these societies. The breakdown was caused by
> attempts, frequently honest and well-meaning, on the part
> of the Western "civilizers" to alter drastically some aspects
> of the way of life of the subject populations without har-
> monizing these alterations with other aspects. The mental
> health and happiness of a people, both as individuals and
> as a group, is often dependent on the possession of a
> coherent culture pattern.

Deep psychological needs of the individual are served by participating in the larger, more complex whole. To be an accepted
part of a group, to be needed and loved by others of our kind—
these are important ingredients for a happy life. Loneliness is
perhaps the most terrible condition, a state which we fear from
childhood to the grave.

It was natural, therefore, that our ancestors began very early
to form communities—first extended families, then tribes, then
larger and more elaborate societies as the advantages developed
throughout history. The societies of primitive man combined
some of the features of the canine and the great ape communities. Early hominids established settlements from which the

males went out to hunt in bands using their stone axes and spears. Cooperation in the planning and execution of the hunt were vital factors in this way of life. When a kill was made, meat was cut from the carcass, brought home, and shared with other members of the tribe. The females cared for the young and became gatherers, picking fruits and nuts and roots which were also carried home and shared. This social structure was better integrated, more highly differentiated, than the family groups of our near relatives—the chimpanzee and the gorilla.

An important biological change had occurred which contributed to the stability of the hominid society. The female of the species no longer went into estrus but was continuously receptive to the male. The excitement and tension that always accompany the monthly periods of estrus for each female are very disruptive factors in other primate communities, as anthropologists have observed in studying the life of the chimpanzee, the gorilla, and the baboon today. Like an unwanted quantum of extra energy cast off by an excited atom, this disruptive factor was removed (or at least greatly reduced) in the hominid society, and this change created a more stable, harmonious community. Long-term relationships were built up between males and females, and the father could be added to the family group. The quieter life pattern of the female allowed for a more consistent nurturing of the young, and as a result their extended childhood did not jeopardize the survival of the species. In fact this lengthened immaturity was turned to advantage, providing a longer period for learning.

The young primate does not come into the world completely equipped with instincts to guide him through all the phases of his life, like many more primitive life forms. Instinct is replaced by education, a time-consuming and less efficient process which allows, however, for large individual differences and is flexible enough to accommodate important creative changes.

Although the ability to learn depends upon the possession of a brain, the amount of brain required for some kinds of association is extremely small. Even a snail can be conditioned by a frequently repeated lesson, as scientists have demonstrated in experiments on the sea mollusk called *Aplysia*, one of the simplest members of the animal kingdom. In the laboratory each snail was touched many times at two different places. The stimulus at one of the sites was routinely accompanied by a mild electric shock. The stimulus at the other was not. Later when the animal was touched at the first site a very vigorous contraction occurred as though the snail anticipated the shock that would follow, while a touch at the second location produced a much less positive reaction.

Just how the lesson is learned and retained by the organism is a question which has not been clearly answered today. Several possibilities are under investigation, but all of them involve the assumption that a change occurs in the chemistry of the system. For example, the information may be stored in molecular form, by an alteration in the shape or arrangement of one of the complex organic molecules. The mechanism may be similar to that involved in the transmission of instinct. The accumulated experience of the species is passed in a code from parent to offspring, and thus behavior patterns are "wired in" from birth. The control of behavior by this method requires a relatively simple type of nervous system with far fewer nerve cells than the multiple switchboards that underlie intelligence. Yet even these rudimentary nervous systems represent a considerable step up in organization from primitive living things like the sponge that have no nervous system at all.

As complex systems evolved, the control of behavior by instinct was gradually replaced by the more sophisticated, more flexible process of learning. The young monkey or ape learns by imitation, watching every move of the adults around him and reinforcing these lessons by practising his skills in play with his siblings and other immature members of the community.

He can also learn to take advantage of a new situation. Jane Goodall gives us a charming example of this ability. A large male chimp who had become quite friendly and trusting (Jane had named him David Graybeard) approached her tent one day when she was working at a table on the verandah. A banana lay on the table beside her. David stood for a while watching Jane. Then suddenly, with one swift but gentle movement he reached out and swept up the banana and loped back to the forest with it. Later Jane placed another banana in the same location. As she had hoped David returned and went right to the table to claim it. On subsequent days the offering and collection of this favorite food was repeated. Gradually other chimps began coming with David and soon this ritual became a general feeding time for the whole troop.

As in the great ape communities, learning by imitation was undoubtedly the principal conditioning of the very primitive hominid young. However, the training was extended through a longer period of immaturity and was probably supplemented to some degree by teaching. The skills needed for the young ape-man were more sophisticated than those demanded of the adult chimpanzee. He had to learn to fashion and utilize a stone axe, to skin an animal, perhaps to lay and tend a fire. As new skills were discovered, communication became more and more important.

To make a society work as an organized whole, there must be transfer of information among the members. So the art of communication developed to fill this need in the evolving organism of the social system. Here again we are dealing with a continuous spectrum of increasing abilities. Methods of communicating by motion or smell are practised by many species, even those with very primitive nervous systems like the bee and the ant. Birds communicate by song. The primates also developed vocal methods of signalling to each other. "The apes have a large vocabulary of calls," Jane Goodall reported, "each signifying an emotion such as fear, pain, or pleasure. When a group arrives at a food-laden tree and gives excited 'food barks' other chimps within earshot often call in response and hurry to the feast. If one chimp gives a low uneasy 'hoo' when he sights an alarming object, other chimps always peer in the same direction. When a youngster screams in fright or pain, his mother invariably hurries to him."

The ability to form words and to speak is, however, a uniquely human accomplishment and seems to depend on the development of certain areas in the brain. The ape lacks these elaborated areas, although there is nothing in the structure of his mouth or larynx that prevents speech. When infant apes are raised along with human babies and given the same training, they do not learn to speak, while the children are soon express-ing themselves fluently. Young humans, even those living in very primitive societies, can learn a language easily.

The word as a symbol for a thing or a situation is probably the key invention that transformed apeman into *Homo sapiens*. Words facilitate the transfer and development of abstract thought. As soon as these mental tools became available they opened up a whole new world of the mind. Like many other skills, language probably began in a very simple way, and the brain evolved to accommodate this new power as it became advantageous. By 40,000 years ago language was in use, and by this time *Homo sapiens* had become established in many areas throughout the world.

With the invention of language the evolution of Form en-tered a new, more dynamic phase. The ability to pass on experi-ence from one generation to another provides a cumulative tradi-tion which is highly adaptive to changing conditions. Through reason, planning, and forethought the process of form-building can move forward more rapidly than by the blind, wasteful, and often cruel method of survival of the fittest. On the other hand, biological evolution was not replaced by the new mode; it has continued to exist along with the human phase just as the evolu-tion of inert matter continued to take place after life entered the cosmos.

But while the new powers of reason were placing increased responsibility and a more active role on the individual, human beings were becoming increasingly dependent on group exis-tence. The small communities of early man were gradually elabo-rating into more populous societies. Swept up in these larger organizations man was beginning to develop the qualities that we recognize as uniquely human—self-sacrifice, altruism, compas-sion. Although the rudiments of these civilizing characteristics were present in societies of species other than our own, they matured and deepened as the social systems became more highly integrated. Far back in what we have assumed to be the savage darkness of prehistoric times there are evidences that love and cooperation with others began to extend beyond the tight bio-logical bonds of male and female, mother and child. It survived after the reproductive period was ended; it lasted after the child had ceased to be dependent; and it extended to many unrelated members of the community. Remains of very ancient settlements prove that the dead were laid to rest with tenderness. The appearance of these traits, as Loren Eiseley suggested, marked a watershed, distinguishing man from beast:

> Here across the millennia we can observe a very moving spectacle, these men whom scientists had contended to possess no thoughts beyond those of the brute, had laid down their dead in grief.
>
> Massive flint-hardened hands had shaped a sepulcher and placed flat stones to guard the dead man's head. A haunch of meat had been left to aid the dead man's journey. Worked flints, a little treasure of the human dawn, had been poured lovingly into the grave. And down the untold centuries the message had come without words: "We too were human, we too suffered, we too believed the grave is not the end. We too, whose faces affright you now, knew human agony and human love." . . . It is the human gesture by which we know a man, though he looks out upon us under a brow reminiscent of the ape . . . he partakes both of Darwinian toughness, resilience, and something else, a humanity . . . that runs well nigh as deep as time itself.

The flowering of these socializing traits is essential to replace the loss of those instincts which control the integration of lower forms of life into smoothly functioning social systems. But our species is very young. The humanizing qualities are still imperfectly developed and unevenly distributed among our kind. Many individuals are more governed by primitive emotions of anger and hatred than by cooperation and concern for others. Mankind is a center of creative ferment where a higher social organism is struggling to emerge from a lower one. In Nietzsche's words, he is "a rope stretched between the animal and the superman—a rope over an abyss."

In the meantime the powers unleashed by the mind continue to evolve at an ever accelerating pace. There is good reason to believe that only a small fraction of the mind's potential has yet been realized; we cannot even guess what powers still lie hidden there. The human brain is the most highly organized assembly of matter that has ever existed on earth—perhaps in the cosmos. The DNA molecule, that masterpiece of the evolution of matter, contains in its simplest form about ten billion atoms, but the mind of man is many times more complex, more detailed, more intricately wrought. Its birth initiated a vastly expedited phase of the evolution of Form. Words set down in writing, coded in patterns of energy, can be instantaneously reproduced, disseminated throughout space and down through time. In this way, the transformation process has been set free of the limitations that have restricted the speed and extent of its spread throughout the cosmos. But the wonderful organ that has made all this possible cannot exist alone. The brain is only one component of a larger whole—a human being with all his inadequacies and inconsistencies.

We have indeed come a long way from the enchanted garden where man was born. The images called up by human history are not the pure, innocent ones of Eden. They present a rapidly changing kaleidoscope of light and shadow. There are dark images of the charred bodies in the furnaces of Auschwitz and on the streets of Nagasaki. But there are luminous images, too: the symmetry of the Parthenon, the poetry of Keats, the delicate tracery of the Taj Mahal, the symphonies of Brahms.

Now in the vast spaces between the stars organic molecules like those that first formed life float on waves of energy that carry messages across the Milky Way. Because man has entered the cosmos, orderly patterns of form have taken wing and are spreading through the universe at the speed of light.

MIND AND ORDER IN THE UNIVERSE

Chaos is but unperceived order; it is a word indicating the limitations of the human mind and the paucity of observational facts. The words 'chaos,' 'accidental,' 'chance,' 'unpredictable,' are conveniences behind which we hide our ignorance.

HARLOW SHAPLEY
Of Stars and Men

There are times when nature appears to revel in chaos and destruction. Breaking abruptly into the reassuring rhythm of quiet days and peaceful nights, a summer storm can suddenly materialize out of a blue and tranquil sky, shattering it like a fine crystal bowl into a thousand shards. Last summer I saw such a storm descend on gentle Ohio countryside. It was late afternoon; the sun, low in the sky, had been flooding the fields with warm evening light, gilding the little pond with its soft fringe of bull-rushes and willow trees. All at once dark cumulus clouds appeared on the western horizon, blocking out the light; the storm descended with an earth-shaking clap of thunder and swiftly turned the landscape into a torn and turbulent scene. Lightning rent open the boiling sky; the air was filled with flying leaves and branches. Sheets of raindrops, driven by wind gusts, fell like daggers on the field of ripening wheat which just a little while before had been rippling beneath the caress of a summer breeze. Clouds of thistledown, torn from the plants, were plastered in gray lumps against the treetrunks and fence posts and windowpanes. At the height of the storm hail as big as robins' eggs fell through the sultry air, piercing the tender leaves of the young corn, flattening the wheat into swirls like giant birds' nests. Soon the ice lay thick upon the fields where soybeans had just begun to draw delicate green lines against the dark earth. In less than twenty minutes the storm had destroyed what sun and gentle rains had nourished into being over many months.

But then the clouds parted. The sun shone through, and above the battered landscape a rainbow flooded the eastern sky with color. It arched above the rim of distant hills, violet shading to blue to green to yellow to orange and finally to deep rose, vibrant against the dark clouds. Higher still, flung against the flickering sky, a pale secondary bow echoed the same colors in reverse: rose, orange, yellow, green, blue, and violet—as though nature had turned a musical scale into color, moving up the scale, taking a deep breath, and moving down again.

Is there any single manifestation of the physical world more ethereal and more symbolic of order than a rainbow? From earliest times mankind has been fascinated and challenged by it. Ancient Peruvians were so awestruck by its appearance that they remained silent until this mysterious sign had faded from the sky. In myths and legends of many cultures the rainbow was believed to be a bridge between heaven and earth—"the pathway of souls"—"the floating bridge to heaven." No one has ever found the pot of gold which legend placed at the end of the rainbow because, as we move toward the bow, it retreats before us. If we turn our backs on it and move away, it follows us like a shadow.

Although ancient man responded to the beauty and the mystery of the rainbow, the order embodied in its precisely repeated and predictable pattern was not explained until the seventeenth century when the French mathematician René Descartes revealed the principle that lies behind it.

The rainbow is like an equation written in the sky. It is visible when an observer is between the sun and a rainshower. If an imaginary line is drawn from the sun through the observer's eyes and extended straight on until it strikes the ground in front of him, all of the raindrops which are at an angle of 40° to 42° from this line reflect the sunlight back to the observer and, therefore, the rainbow describes an arc around this axis. The rays passing through the raindrop, reflected from the front surface and passed back again through the drop, are bent through slightly different angles depending on their wavelengths. The red, bent the least, is seen to be highest in the sky at 42°. Then the others follow in sequence, ending in violet which is refracted through the largest angle. The secondary bow is caused by a double reflection and so the sequence of colors is reversed.

The rainbow would be a perfect circle if it were not cut off by the earth. Under unusual circumstances, when the sun is low and the observer is very high, as in an airplane flying between the sun and a raincloud, the complete circle of the rainbow can be seen with the shadow of the plane impaled like a dark moth in the center of the circle.

Descartes's rational explanation of the rainbow was based on the assumption that is common to all scientific thought: there is an order in nature which can be understood by the human mind. We in the Western world take this assumption for granted because it has long been a part of our cultural heritage. But the thought is foreign to many civilizations. The belief is more commonly held that unfathomable mystery lies at the base of things, that magic and ritual are the most effective ways to deal with the vagaries of nature. Even in great cultures where art and literature have flowered—in India, in Egypt, in Persia—no consistent effort was made to discover an order lying beneath the variety of the external world. And science did not develop in these cultures. Although facts were noted and collected, the leap from the specific to the general was not taken.

It was the Greek philosophers who made this important contribution to human thought. They believed that nature is basically lawful, that an elegant simplicity is concealed behind the transitory external appearances of things. Our word *cosmos* comes from the Greek *kosmos,* meaning an orderly whole. Thus science has been called "thinking about the world in a Greek way." That is why science has never existed except among peoples who came under the influence of Greece. If the Persians had conquered the Greeks science might never have flourished in Europe.

The hypothesis that there is a common substratum beneath the almost infinite variety of shapes, colors, and textures requires an enormous leap of the imagination. The idea that objects as unlike as iron and sand, water and stone, could be different forms of the same fundamental substance was first suggested by Thales of Miletus six centuries before Christ. "All things are made of water," he declared. A century later Democritus proposed a different theory: "The world is composed of indivisible and identical a-tomic grains," he said. "All things are numbers," countered Pythagoras; "the *kosmos* is based on melody."

And so the search began—the search that has led to quarks and quantum theory, to black holes and the fourth dimension. Instead of simplicity science has found complexity; the single fundamental substance has not yet been discovered, but the faith that there is some kind of common denominator in the apparent diversity of nature has not wavered. Without this faith science would never have been born or come of age. In the words of Alfred North Whitehead:

> There can be no living science unless there is a widespread instinctive conviction in the existence of an *Order of Things,* and, in particular, of an *Order of Nature.* I have used the word *instinctive* advisedly. It does not matter what men say in words, so long as their activities are controlled by settled instincts. The words may ultimately

destroy the instincts. But until this has occurred, words do not count. This remark is important in respect to the history of scientific thought. . . .

Of course we all share in this faith, and we therefore believe that the reason for the faith is our apprehension of its truth. But the formation of a general idea—such as the idea of the Order of Nature—and the grasp of its importance, and the observation of its exemplification in a variety of occasions are by no means the necessary consequences of the truth of the idea in question. . . .

The main recurrences of life are too insistent to escape the notice of the least rational of humans; and even before the dawn of rationality, they have impressed themselves upon the instincts of animals. It is unnecessary to labour the point, that in broad outline certain general states of nature recur, and that our very natures have adapted themselves to such repetitions.

But there is a complementary fact which is equally true and equally obvious: nothing ever really recurs in exact detail. No two days are identical, no two winters. What has gone, has gone forever. Accordingly the practical philosophy of mankind has been to expect the broad recurrences, and to accept the details as emanating from the inscrutable womb of things beyond the ken of rationality. Men expected the sun to rise, but the wind bloweth where it listeth. . . .

Faith in reason is the trust that the ultimate natures of things lie together in a harmony which excludes mere arbitrariness. It is the faith that at the base of things we shall not find mere arbitrary mystery.

Follow any of the great scientists through their daily rounds and one will discover that they act upon the instinctive conviction that the multiplicity of appearances can be reduced to a few rational and simple principles. Whenever they find confirmation of this belief they are excited and moved by the discovery. It is an aesthetic experience, as Isaac Newton found in his early experiments on light:

> In the year 1666 I procured me a triangular glass prism to try therewith the celebrated phaenomena of colours.

And in order thereto, having darkened my chamber, and made a small hole in my window-shuts, to let in a convenient quantity of the sun's light, I placed my prism at its entrance, that it might be thereby refracted to the opposite wall. It was at first a very pleasing divertissement, to view the vivid and intense colours produced thereby. . . . But the most surprising, and wonderful composition was that of whiteness. There is no one sort of rays which alone can exhibit this. 'Tis ever compounded, and to its composition, are requisite all the aforesaid primary colours, mixed in a due proportion. I have often with admiration beheld that all the colours of the prism being made to converge, and thereby to be again mixed, as they were in the light before it was incident upon the prism, reproduced light, entirely and perfectly white, and not at all sensibly differing from a direct light of the sun. . . .

The use of scientific method has proved to be the best way to gain a fuller understanding of the world, and through this understanding to achieve control over it. The method involves the meticulous observation of fact, the discovery of orderly sequences in nature and hidden similarities between objects as seemingly unrelated as a crystal and a star, an atom and the solar system, a falling apple and the moon's path across the night sky. Finally science involves the construction of a theory that provides a rational basis for these sequences and similarities. This process, based upon the assumption of an order in nature, has been spectacularly successful and the success is in itself a testimony to the validity of the assumption.

It is oddly ironical, therefore, that the discovery of certain scientific facts began to cast doubt on the belief that order lies at the base of the physical world. Heisenberg's uncertainty principle, as interpreted by some scientists and philosophers, was taken as proof that strict causality does not exist. If we cannot identify cause and effect relationships at the level of fundamental particles, then we must assume that all the order we have found in nature is derived simply from statistical analysis of random events (a fallacy that we examined in some detail in Chapter 7). "It appears," said Erwin Schrödinger, "that there are two different 'mechanisms' by which orderly events can be produced: the 'statistical mechanism' which produces 'order from disorder' and the [other] one producing 'order from order.' To the unprejudiced mind the second principle appears to be much simpler, much more plausible. No doubt it is. That is why physicists were so proud to have fallen in with the other one, the 'order from disorder' principle."

This revolutionary thought caused a reaction throughout the Western intellectual world. Philosophers, poets, and artists were deeply affected by it, but it is an interesting fact that scientists, while paying lip service to the "order from disorder" principle, have continued to pursue their research on the assumption that nature is orderly. When they close the door of their laboratories at the end of a day, they confidently expect that the experiments they have just completed can be repeated tomorrow and that, within certain known limitations of error, the results will be comparable. The same laws of interaction will apply tomorrow just as they did today, and if the field of research is expanded to encompass a larger portion of the physical world, scientists assume that the same principles will continue to apply. The laws of chemical reaction on the moon, on Mars, on Jupiter will be identical to those on Earth. As Whitehead perceived, their activities are guided by the instinctive—or intuitive—conviction that there is an order in nature. What they say in words does not matter.

Most scientists are specialists, dealing with very narrow segments of the physical world, and they do not regularly concern themselves with broad philosophical problems. If asked about the "order from disorder" principle, the majority of researchers would say, "The issue is not relevant to my work." In fact scientists have been trained to view with distrust the introduction of philosophical arguments into science. "That is not physics; that is metaphysics," they declare. And in the meantime they go quietly and effectively about their task of discovering the orderly relationships in nature and reducing these to laws that can be expressed in mathematical equations.

Mathematics is an outstanding example of the ability of the human mind to abstract from the specific to the general—to dis-

The discovery of hidden similarities in nature.

Plate 15
Chambered Nautilus Shell (X-ray photograph, courtesy of Eastman Kodak Company).

Plate 16
Sword Fern (Photograph © Pat O'Hara).

cover basic forms and common attributes in the diverse objects of the physical world. Greek philosophers believed that this process of abstraction was the best way of achieving an understanding of nature. Euclidean geometry, one of the supreme achievements of Greek thought, represents a bold extrapolation from sensory impressions. How often do we encounter a precise right angle in nature? Or a pure circle? Or a perfect helix? Close approximations to these shapes do exist, as we can see now with the aid of powerful microscopes, special photographic methods, and computer technology (see Plates 11 & 15). But in the world of common experience these shapes are usually overlaid with other forms in complex combinations that cannot be unraveled without the use of sophisticated tools and higher mathematics.

The process of abstraction invented by Pythagoras and Euclid created a powerful tool for studying nature. By providing rules for manipulating shapes and forms, geometry revealed orderly relationships and suggested new combinations that led to broader generalizations. Descartes used these principles in explaining the order that lies behind the rainbow. He also found in Euclidean geometry a method for defining the three dimensions of space and invented the Cartesian coordinates. Einstein used this insight to create a more comprehensive synthesis. The Cartesian coordinates were the frames of reference for his Special Theory of Relativity. As it has become more sophisticated, mathematics has given scientists ever more powerful methods for recognizing and defining relationships. In so doing, it has made important contributions to the search for an underlying order in nature.

Just in the last few years technologies based on the use of computers have opened up new ways of understanding structure and form. Fractals, for example—the name comes from the Latin *fractus,* meaning broken or fractured—are patterns constructed from geometric shapes. As in similar triangles, the shape is the constant factor but the scale is varied. In fractals new smaller units are added to the figure, which becomes more complicated with each addition. A classic example, the Koch curve, starts with an equilateral triangle. When smaller equilateral triangles are erected, centered on each side, the figure becomes a six-pointed star. And as even smaller triangles are added in the same manner, the pattern takes on the shape of a more and more elaborate snowflake. In theory, at least, the complexity can be infinitely increased. Using computers to do the time-consuming repetitive work, forms resembling flowers and leaves and even landscapes can be created.

Mathematicians working with this new technology have found that it offers a powerful way of analyzing spatial and temporal phenomena that are too complicated to understand by traditional methods. An underlying pattern can be identified in the structure of large protein molecules, in the movement of undersea currents, and in the flood levels of the Nile river over a thousand years. A common element has been found in all music from the Beatles to Beethoven. As the fractal images take shape on the computer screen the first reaction of the operator is "invariably a kind of intoxication." The instantaneous perception of form in what appeared to be formless confirms the faith that chaos is but unperceived order.

There is no doubt that the human mind prefers order and simplicity. Mere collections of unrelated facts serve no purpose; they must be put together in a way that reveals regularities within them or they cannot be understood. This point was dramatized by Karl Popper, the Austrian philosopher of science. Suppose, he said, that someone decided to devote his whole life to science and he recorded in detail every fact that came to his attention throughout every subsequent day of his life. He filled notebook after notebook with detailed observations: the temperature and humidity, the racing results, the level of cosmic radiation, the stock market quotations, the aspects of the planet Mars. Let us suppose he was so meticulous that nothing was omitted from this record. And when he died he left his notebooks to the Royal Society of London, convinced that they represented a great contribution to science. But the notebooks were useless; they would never be used by the members of the Royal Society because they contained a meaningless jumble of undigested facts.

In their search for order and unity, scientists have needed to make many simplifying assumptions. The very act of reducing objects to geometric shapes is a simplification. Furthermore, it sometimes happens that the same set of facts can be explained in two or more different ways; when this happens the simplest explanation is preferred. A principle known as "Ockham's razor" (proposed by William of Ockham in the fourteenth century) says that it is unsound to set up more than one hypothesis to explain a phenomenon when one will suffice. And this principle has been respected throughout scientific history. The significance and elegance of a scientific theory are measured by its simplicity and the degree to which it makes sense out of what appeared to be unrelated and disorderly facts.

It has been argued that in searching for unity and making simplifying assumptions scientists have imposed order on nature—that the order we think we have found has no objective reality; it is only a reflection of our own minds. As Arthur Eddington expressed it: "We have found a strange footprint on the shores of the unknown. We have devised profound theories, one after another, to account for its origin. At last, we have succeeded in reconstructing the creature that made the footprint. And lo! It is our own."

Intriguing as this thought may be, it does not stand up to logical analysis. Descartes did not invent the order in the rainbow. Newton did not create the spectrum of colors that are concealed in a beam of sunlight. Nor—to choose another example from the history of science—did Johannes Kepler impose the elliptical shape on the orbit of Mars. Although Kepler would have liked to find a simple geometric relationship in the orbits of the five known planets, and in fact did try to do this without success, he finally discovered the true shape of Mars's orbit after seven years of laborious cut-and-try calculations. His original work, still preserved, covers nine hundred folio pages in small handwriting. The story of the way this discovery was made is illustrative of the spirit and method of all good scientific work.

In the year 1600 Kepler was assigned the task of defining the shape of the orbit of Mars, using the observations of Tycho Brahe, the most precise astronomical observations that had been made up until that time. It was obvious from these measurements that Mars did not move in a perfect circle around its center in the sun, as natural philosophers had long assumed. But Kepler did not immediately reject the idea of a circle—the simplest and most beautiful geometric shape. He started with the hypothesis that the planet moved in a circular orbit with the sun displaced from the center. After five years of calculation he found a solution that fitted the measurements very closely, so closely in fact that the largest deviation from this circular path that had been observed was eight minutes of arc. This is an angle so small that it would subtend only the width of a pencil line if the angle were drawn between your eye and this page. But even this tiny deviation could not be ignored. It was slightly larger than the margin of error in Brahe's measurements; and facts this well established, Kepler said, must not be coerced to fit a theory. Renouncing his earlier conclusion, Kepler spent two more years in computation, trying many different arrangements of sun and planet, before he stumbled upon the geometric shape that fitted perfectly with the known observations. The orbit of Mars is an ellipse with the sun at one focus.

This combination of rigorous respect for accurate observation with a passionate interest in the discovery of unifying principles is characteristic of Western science, and has made the modern world. "It determined the climate of European thought in the last three centuries," observed Arthur Koestler, "it set modern Europe apart from all other civilizations in the past and present, and enabled it to transform its natural and social environment as completely as if a new species had arisen on this planet." The success of the method strongly supports the validity of the assumption upon which it is based.

But there is another, even more important reason for rejecting the notion that the order we find in nature is just a reflection of our own minds. This notion rests upon the belief that man is something different from the rest of the universe—built of different world-stuff, moved by different forces. The attitude is a legacy from the religious doctrine that set man apart from the rest of creation and gave him dominion over nature. The new view sees man as the present apex of a long-sustained creative process encompassing the entire cosmos. The order that we find in nature is a characteristic of both nature and the human mind because man is an integral part of nature.

Mind is the present culmination of traits that have been building throughout evolutionary history: greater awareness of the rest of the physical world and increased powers to alter and control it. The ability to identify orderly elements in nature enabled primitive man and many simpler organisms to deal with life situations more effectively. By recognizing and anticipating regular changes—the rise and fall of temperatures with the seasons, the alternating abundance and scarcity of food—they were able to develop survival techniques, find shelter, and lay away stores of provender for the winter. Thus natural selection encouraged the development of a nervous system that could sort like elements out of diverse sensory impressions. The orderly processes that evolved in the brain matched reality, and therefore were useful. Rationality could not have evolved amid chaos; it would have had no survival value.

Survival value alone, however, is not sufficient to explain the importance of order to all living things. The effectiveness of the whole organism depends upon finding orderly elements in the environment. This is true far down the evolutionary scale, as research with laboratory animals has demonstrated. In one experiment rats were placed in a cage with a door that could be pushed open to find food, and they quickly became accustomed to finding food in this predictable way. Then the mechanism was changed so it operated in a random manner. After a period of this treatment, even though sufficient nourishment was provided, the animals became listless and lost their will to live.

Human beings also care about form and order in ways that go well beyond the physical needs driven by natural selection. They are moved and delighted by the sight of ice crystals spreading across the surface of a pond on a winter day, or by the regular cadence of long rollers breaking on a sandy shore. They are awed by the pure, translucent geometry of a scallop shell or the fine-scaled structure of a dragonfly's wing. Many people spend their whole lives creating harmonious arrangements of color and pattern, of melody and tone, of surfaces and space. These enthusiasms and drives cannot be explained on the basis of usefulness or the survival of the fittest.

Alfred Russel Wallace, who, simultaneously with Charles Darwin, proposed the theory of natural selection, expressed serious doubts that selection could have produced the mental abilities that have nothing to do with biological fitness, such as the ability to compose great music, to design artistic creations, and devise new mathematical systems. Wallace outlined this position in an article which he sent to Charles Darwin. It has been reported that Darwin was deeply distressed when he read Wallace's article and wrote "No" . . . "No" . . . "No"—heavily underlined across the paper. But Darwin was not able to muster effective arguments against Wallace's conclusion.

The questions raised by Wallace can be answered in the light of the theory of evolving Form. Physical survival is only one of many ways in which Form can be extended. Each higher level of complexity opens up new possibilities for the creation of order. Once developed, the organizing power of the brain can be used in other formative ways to discover and to create patterns of sound, shape, and color. These activities are not related to self-preservation; they represent another expression of the transformation process.

Art, music, poetry—all the aesthetic works of man—are successful to the extent that they find or create order where there was none before. The measured cadence of a poem, the rhythmic repetition of columns in a Greek temple, the recurring theme in a symphony—all these creations of form satisfy a deep need which is built into our irrational as well as our rational nature. "A series of random musical notes means nothing to us," remarked Edmund Sinnott, "but when they are arranged in a particular order and rhythm, as in the opening bars of Beethoven's Fifth Symphony, they stir our emotions. A random spattering of pigments on a canvas means little, but if these have a particular arrangement, a masterpiece emerges. Beauty is orderly, not chaotic. It is an organized pattern of sights or sounds or words or images which strikes a chord within us; which vibrates, so to speak, on our particular wavelength."

The sympathetic chord which beauty strikes within us is the

apprehension of order in the external world, reinforcing the intuitive perception of the unity of man and nature. The painter and the sculptor reveal the underlying shapes and patterns in the multiplicity of visual experience. A Cezanne still life brings out the roundness of an apple, the tapered curve of a pear. Brancusi's *Bird in Space* expresses in three dimensions the smooth fluid line of a bird's flight. The poet, like the scientist, explores unexpected similarities: "The fog comes / on little cat feet"; ". . . the sunken / tide-rocks lift streaming shoulders / out of the slack"; "the white mares of the moon rush along the sky, / beating their golden hoofs upon the glass Heavens."

Something central to our being is satisfied by the recognition that all things exist together in harmony. When we see a full moon rise over the sea, or stand beneath the soaring arches of a great cathedral, or sit in a symphony hall transported on the wings of sound, tranquility replaces tension and confusion. Each experience of beauty reaffirms our faith that form and order lie at the heart of the universe.

It is one of the saddest facts of modern times that this most basic need has been seriously undermined by the twentieth century scientific world-view. Ideas lifted from science (and attributed a greater degree of certainty than they possess) have been transplanted into the sensitive areas of the humanities. Here they have caused confusion, frustration, and cynicism. It is considered sophisticated to accept gracefully the "scientific fact" that the universe has no purpose and our own lives are without meaning. Like caged rats suddenly precipitated into a disorderly environment that cannot be understood, we see ourselves as rational beings cast adrift in an irrational universe. With such ideas underlying modern thought it is no wonder that a sense of futility and despair—even a loss of the will to live—have struck at the heart of man. These negative feelings have found expression in twentieth century art, music, even politics and our dealings with our own kind. "It was already obvious by the beginning of this century," said Archibald MacLeish, "that many of our artists and writers—those not so silent observers of the human world who sit in its windows and lurk in its doorways watching—were not precisely in love with the modern world, were, indeed, so little in love with it that they had turned against life itself, accepting absurdity and terror in its place and making of human hopelessness the only human hope."

Terror is made visual in the nightmarish swirls of Edvard Munch's *The Scream;* it is slashed in black across Georges Rouault's *Head of Christ.* Absurdity and human hopelessness are eloquently expressed in Franz Kafka's *Metamorphosis,* depicting the transformation of a man into a monstrous cockroach. "Vermin," Kafka said, "are born of nothingness." A similar theme is developed by Eugene Ionesco in *Rhinoceros,* a drama in which men, unable to maintain their integrity and identity, turn into beasts. And Samuel Beckett, in *Waiting for Godot,* portrays with compassion the deferral of hope—the meaninglessness of life: "The tears of the world are a constant quantity. For each one who begins to weep somewhere else another stops. The same is true of the laugh. Let us not then speak ill of our generation, it is not any unhappier than its predecessors. Let us not speak well of it either. Let us not speak of it at all."

Scientists have been held responsible for the malaise that has overtaken mankind. They are believed to have discovered that the world is purposeless, that dreams of a meaningful universe must be laid aside, and that mankind must learn to live honorably in a bleak and sterile world. This thought was movingly expressed by Bertrand Russell:

> Such, in outline, but even more purposeless, more void of meaning, is the world which Science presents for our belief. Amid such a world, if anywhere, our ideals henceforward must find a home. That man is the product of causes which had no prevision of the end they were achieving; that his origin, his growth, his hopes and fears, his loves and his beliefs, are but the outcome of accidental collocations of atoms; that no fire, no heroism, no intensity of thought and feeling, can preserve an individual life beyond the grave; that all the labours of the ages, all the devotion, all the inspirations, all the noonday brightness of human genius, are destined to extinction in the vast death of the solar system, and that the whole temple

of Man's achievement must inevitably be buried beneath the debris of a universe in ruins—all these things, if not quite beyond dispute, are yet so nearly certain, that no philosophy which rejects them can hope to stand. Only within the scaffolding of these truths, only on the firm foundation of unyielding despair, can the soul's habitation henceforth be safely built. . . .

Brief and powerless is Man's life; on him and all his race the slow, sure doom falls pitiless and dark. Blind to good and evil, reckless of destruction, omnipotent matter rolls on its relentless way; for Man, condemned to-day to lose his dearest, to-morrow himself to pass through the gate of darkness, it remains only to cherish, ere yet the blow falls, the lofty thoughts that ennoble his little day; disdaining the coward terrors of the slave of Fate, to worship at the shrine that his own hands have built; undismayed by the empire of chance, to preserve a mind free from the wanton tyranny that rules his outward life; proudly defiant of the irresistible forces that tolerate, for a moment, his knowledge and his condemnation, to sustain alone, a weary but unyielding Atlas, the world that his own ideals have fashioned despite the trampling march of unconscious power.

Although Russell recognized that the conclusions on which this philosophy was based are "not quite beyond dispute," he assigned them too great a degree of certainty. Unlike religious revelations, the scientific interpretation of underlying reality is always provisional. The ultimate truth about the universe has not been revealed by science and perhaps never will be, although it is constantly moving toward that goal. As knowledge and understanding grow, well-accepted theories frequently give way to more comprehensive conceptual schemes, and these result in quite different world-views. We have seen dramatic reversals in thought just in the last hundred years, and change is in the air again. The concept of wholeness, spearheaded by the science of biology, is receiving more and more serious attention, not only in the field of biology but in many other sciences as well. We encounter it in the work of geneticists, anthropologists, biochemists, and even astronomers. The success of the analytical method has nattural limitations, and these limitations are being approached in many scientific fields. Researchers are beginning to recognize that systems must be treated as single units—that they are more than the sum of their parts. This is an idea whose time has come.

The theory proposed in this book carries the concept of organism several steps further with the recognition that self-organized units, both living and inert, are different from artificially imposed arrangements of matter. They are not passively pushed and pulled by external influences. They possess inner resources, and they act to preserve and extend themselves in space-time. Each organism is a center of creativity and throughout time more and more intricately designed organisms have made their appearance, culminating in mankind—the present apex of the creative process.

These are positive concepts that restore meaning to the cosmos and are capable of bringing about a resurgence of belief in ourselves. The importance of a strong self-image cannot be overestimated, because in human affairs an idea is a greater moving force than any physical influence. Throughout history men and women have died for abstract principles. They have turned themselves into living torches, allowed themselves to be torn apart by wild beasts or skinned alive rather than deny their beliefs. So the shape of the future will depend to a large extent on our understanding of our role in the cosmic process. "What man believes about *himself* is of utmost moment," said Edmund Sinnott, "for it will determine the kind of world he will make and even his own fate."

IMPLICATIONS FOR THE FUTURE OF MAN

*Our part in the universe may possibly in some
distant way be analogous to that of the cells in an
organized body, and our personalities may be the
transient but essential elements of an immortal
and cosmic mind.*

FRANCIS GALTON
*Inquiries into Human Faculty
and Its Development*

The lines quoted above were written in 1883 by a man whose
intellectual imagination placed him at least a century ahead of
his time. Grandson of Erasmus Darwin, first cousin to Charles
Darwin, Galton is best known as the founder of the science of
eugenics. The idea that mankind could be improved by genetic
control is still considered a radical proposal today. In another
revolutionary insight Galton recognized the remarkable phenom-

enon of whole-making and the mysterious relationship between the whole and its parts. "We as yet understand nothing," he wrote, "of the way in which our conscious selves are related to the separate lives of the billions of cells of which the body of each of us is composed. We only know that the cells form a vast nation, some members of which are always dying and others growing to supply their place, and that the continual sequence of these multitudes of little lives has its outcome in the larger and conscious life of the man as a whole." Galton suggested that mankind may become part of an even more important whole— the cosmic mind. This is a thought that is just beginning to take shape in some of the most forward-looking philosophies of the present day.

Is mankind evolving into a more highly integrated whole? The theory we have been developing leads logically to this conclusion. We have seen that individual units tend to combine to form larger wholes, building more and more intricate, more finely differentiated systems. But at the same time there is also a tendency for each individual organism to become more complex, more capable, more aware of itself, and endowed with more freedom of choice. These two trends appear at first sight to be contradictory, but as we look at the process more carefully we find that they are complementary.

In general, when organisms unite to form a larger whole, the structure of the individual units is not sacrificed. The whole is absolutely dependent on its parts but so, too, are the parts enhanced by being swept up in the whole. Oneness and separateness exist harmoniously side by side. Electrons can enter and leave an atom as single identical units of matter-energy; when a crystal is melted the molecules that composed it disperse again as separate units; the sponge can be broken down into many independent living cells; termites retain their own existence even while participating in an elaborately organized society. Even in an organism as complex as a human being, an integrated system of a hundred trillion cells, some of these cells can be removed and—given the proper nourishment and environmental conditions—they live, grow, and reproduce themselves.

There are indeed a few cases that seem to be exceptions to this general rule—for example the slime fungus which belongs to a group technically known as *Myxomycetes*. This strangely formless creature is an unpleasant-looking object that usually causes a shudder of distaste when it is discovered creeping along decaying logs or piles of rotting leaves. A thin, slimy film of protoplasm without enduring shape or structure, it is not divided into separate cells and has no specialized parts. But even this odd shapeless organism undergoes a reproductive phase and gives rise—miraculously it would seem—to thousands of individual spores which are distributed by the wind. Under favorable circumstances these spores sprout into tiny duplicates of the parent. The new organisms grow by adding to their bodies other individuals of the same species. Whenever individual units encounter each other they join eagerly together, completely blending into one gelatinous mass. Thus the large organism is in reality a union of hundreds of spores that have entirely sacrificed their separateness. This act of total coalescence is so foreign to our common experience it is difficult to visualize. "Imagine," said H. G. Wells, "that when two people meet each other on the street they run together into one blob, as drops of water run together, so that ultimately the whole population of a town is rolled up into a gigantic mass of living substance that creeps about like a single creature."

Fortunately there is little danger that this unattractive destiny awaits our species. The type of synthesis exemplified by this slime fungus appears to be an evolutionary dead end. The species organized in this manner have not radiated or developed into more advanced forms of life. In fact, relinquishing of individual structure is very rare in nature. Structure is usually retained as each new level is erected on the base of the one that went before. Like a skyscraper made up of individual rooms, the form of each unit must remain intact when the assembly is built or the entire edifice will come tumbling down. In an organism, as Galton observed, the multitudinous little lives are preserved even as the larger whole is created.

If we project the course of this transformation process into the future, we would expect that any superorganism in which mankind participates will retain the integrity of individuality, while at the same time the new whole will be a uniquely endowed creation. Observing mankind from this perspective, we can see that several aspects of collective organization are already well advanced. Human societies have developed rapidly into larger, more finely differentiated systems. And, like a living organism, they exhibit the properties of growth and specialization of parts. As Herbert Spencer has remarked:

> It is a character of social bodies, as of living bodies, that while they increase in size they increase in structure. A low animal, or the embryo of a high one, has few distinguishable parts; but along with its acquirement of greater mass, its parts multiply and simultaneously differentiate. It is thus with a society. . . . The various groups have various occupations. . . . This division of labour, first dwelt on by political economists as a social phenomenon, and thereupon recognized by biologists as a phenomenon of living bodies, which they called the 'physiological division of labour,' is that which in the society, as in an animal, makes it a living whole. Scarcely can I emphasize sufficiently the truth that in respect of this fundamental trait, a social organism and an individual organism are entirely alike.

Some forms of government, Spencer suggested, may become too rigid, resulting in an increased subordination of the individual. In such cases, the evolution of greater complexity might produce a community of human ants or bees. This analogy has been mentioned by other writers as well. René Dubos said: "The large cities of the modern world remind one of beehives and of ant hills. Each individual person in them has a specialized function and returns to rest at a particular place—as if he were but one among so many other interchangeable units in an immense colony. As human societies become larger, older, and more dependent on technology, the colonial organization becomes more intricate and less flexible. The formal resemblance between human institutions and the colonies of social insects is indeed so striking that one might conclude that modern man is doomed to become a specialized worker or soldier in a stereotyped social machine."

However, Dubos then goes on to point out an important difference between these two kinds of society. In the insect colonies there is close genetic relationship between all the individuals. They are members of one extended family, while in human societies there is great genetic diversity.

In other ways, too, there are significant differences between the human and the arthropod societies. Each individual insect is automatically controlled, programmed by instinct, and thus it is endowed with a lower degree of freedom. Under normal circumstances the individual automatically does what is best for him and for his species.

In the human phase of evolution new factors have entered into the process. Cooperative behavior is voluntary rather than chemically controlled, yet it is a dominant force in human relations. The presence of such socializing traits as altruism and concern for others is a fact that cannot be explained on the basis of old biological imperatives. The emergence of these unifying qualities can best be explained as a stage in the form-building process. They are the direct result of the need to participate in a larger, more harmoniously self-organized whole—a product of the inner force that created the helium atom, the ice crystal, the living cell. But the units that make up human societies each possess capabilities far beyond those of any other organisms. The larger whole built on these units will necessarily contain elements that were not present at any previous stage of the creative process.

One of these new elements is the invention of efficient modes of communication. The effective organization of each system depends upon a rapid and reliable method of transferring messages among the parts that comprise the whole. In many organisms the parts are in actual contact, so signals coded in electrical impulses or particles of matter can be passed directly from one part to another. In some situations, however, the units are not contiguous, and then the information must travel through space. One striking example of this type of organization is seen in schooling fish.

Hundreds of individuals move together as though they were part of a single body, wheeling and turning in perfect synchronization. Information passes between the fish by sight and by sound, which travels very rapidly in water. Mankind has invented methods that overcome spatial limitations to a far greater degree. We can speak with only a fraction of a second delay to friends in Australia, or hold a business conference call between Japan, Switzerland, and New York.

More significantly, however, we have discovered ways of overcoming the limitations of time—an accomplishment that no other organism has achieved even in a very limited way. The ability to store and pass on knowledge and experience has produced a cultural tradition, a factor that has appeared only once in evolutionary history. This tradition grows and undergoes adaptation. Its evolution has added another layer to the transformation process, superseding—although not replacing—the old biological evolution, and moving at a much faster pace, because these elements are more easily modified than physical traits.

Transmission of the written word, speeded first by printing, and more recently by the invention of the magnetic tape, has made possible the building up of a suprahuman memory—the collective intellectual life of mankind, extending back to the dawn of civilization. Stored in libraries and museums around the world, this memory can be tapped by every individual, who may add his personal interpretation or perhaps an additional contribution of knowledge to it. Each generation inherits a larger cultural tradition than the one that went before; so each generation "stands on the shoulders of giants," and can—in theory at least—accomplish more because it starts from a broader base of knowledge. Although the individual passes away, his thoughts become part of this rapidly augmented, undying flood of thought. By participating in this greater whole the individual overcomes the briefness of his own existence. As H. G. Wells expressed it: "He escapes from his ego by this merger, and acquires an impersonal immortality in the association; his identity dissolving into the greater identity. This is the essence of much religious mysticism, and it is remarkable how closely the biological analysis of individuality brings us to the mystics. The individual, according to this line of thought, saves himself by losing himself."

In a very real and exciting way the individual human being has attained an important role in the creative process. In the biological phase of evolution, genetic information passes directly to succeeding generations, bypassing the individual; so the effort of a single organism cannot alter the genotype or influence the development of the species. But through cultural evolution each person can influence the future. For the first time in history a single organism is capable of affecting the evolution of Form.

Having achieved these remarkable new powers, it is no wonder that people look with distrust on any theory that implies the subjugation of the individual to a larger whole. The "organization man" is a concept that raises many specters—the sacrifice of autonomy, reduction to a mechanical role in a vast rigid society. But these fears are based on false models: the honeybee colony, the termite society, or—the ultimate horror—the shapeless sheet of slime fungi. As we have seen, the future of mankind cannot be projected by reference to these earlier modes of organization. The creative potential of the individual has been meticulously built by the evolutionary process over billions of years, and it need not be sacrificed as the next stage comes into being. "What is the work of works for man," Teilhard de Chardin asked, "if not to establish, in and by each one of us, an absolutely original center in which the universe reflects itself in a unique and inimitable way?"

This description is reminiscent of the relationships revealed in the fractal, where the shape of the complete pattern is identical with the shape of each portion, no matter how small. Self-symmetry is present in ever-diminishing scale, and the significance of each tiny unit is multiplied manyfold by its participation in the total pattern.

Even the most cautious projections of the transformation process to the next stage must be recognized as tentative extrapolations from present knowledge. But the implications are too important to be passed over without comment. Visions of future possibilities cast light upon the present, help to define a sense of direction, and provide a new evaluation of trends that are already evident today.

It has been suggested that the next development in the creative process may be the formation of a global or even a cosmic mind. Teilhard de Chardin expressed the thought this way:

> We are faced with a harmonized collectivity of consciousnesses equivalent to a sort of super-consciousness. The idea is that of the earth not only becoming covered by myriads of grains of thought, but becoming enclosed in a single thinking envelope so as to form, functionally, no more than a single vast grain of thought on the sidereal scale, the plurality of individual reflections grouping themselves together and reinforcing one another in the act of a single unanimous reflection.
>
> This is the general form in which, by analogy and in symmetry with the past, we are led scientifically to envisage the future of mankind, without whom no terrestrial issue is open to the terrestrial demands of our action.
>
> To the common sense of the "man in the street" and even to a certain philosophy of the world to which nothing is possible save what has always been, perspectives such as these will seem highly improbable. But to a mind become familiar with the fantastic dimensions of the universe they will, on the contrary, seem quite natural, because they are directly proportionate with astronomical immensities.
>
> In the direction of thought, could the universe terminate with anything less than the measureless—any more than it could in the direction of time and space?

Teilhard's visionary thoughts are clothed in beautiful but obscure analogies. What is really meant by a single thinking envelope? a vast grain of thought on the sidereal scale? Stripping the concept of some of its mystery, we can imagine that minds communicating with other minds around the world could be visualized as single cells which together form a collective mind, a mind that is much more capable than a single brain functioning alone. Development in this direction is apparent at the present time. Megasynthesis of intellectual activity is taking place throughout the civilized world and is gathering momentum, although individual human beings are not conscious of relinquishing any sovereignty as higher degrees of interaction are achieved. The means of relating with each other—the "connectivity," as it has been called—are approaching a very sophisticated stage, in which individual human beings, still retaining their separate personalities, are acting as integral parts of a larger intellectual system. For example, the whole fabric of contemporary science is a single collective endeavor made possible by publication and the establishment of an international community of scientists, all working to extend the limits of knowledge. Together they achieve results far beyond the abilities of the single scientist working alone.

Although science has achieved a higher level of world-wide integration than most areas of human thought, other disciplines, too, are beginning to follow the same trend. As each works out its own methods of interaction and cross-fertilization among all the individuals involved in that field of thought, they are helping to build a cohesive world culture.

Furthermore, there is some evidence that a synthesis of minds can produce results that transcend mere exchange of information among separate individuals. One interesting illustration of this point is the phenomenon of simultaneous discovery. In 1666 young Isaac Newton invented the differential calculus; a few years later the same principles were worked out independently by Gottfried Wilhelm von Leibnitz in Germany. In 1859 the theory of natural selection as the mechanism for evolution was published simultaneously by Charles Darwin writing in England and Alfred Russel Wallace in the Malay Archipelago. These and many other well-documented cases are consequences of the developing collective mentality of mankind. Knowledge and understanding within the field in question had reached a critical level of sophistication. The time was ripe for these discoveries—the ideas were in the air. They were, perhaps, manifestations of a synthesis that holds immeasurably greater potential.

The concept of an evolving global mind is an idea that is in the air today. Lewis Thomas said: "It seems to me a good guess, hazarded by a good many people who have thought about it, that we may be engaged in the formation of something like a mind for the life of this planet."

British psychologist Peter Russell developed this thought further in his recent book *The Global Brain*. Building on James Lovelock's theory of Gaia (the earth as a whole shows characteristics of organization and self-regulation), Russell suggested that if mankind were to evolve into a healthy, integrated social superorganism, this transformation could create a nervous system for the earth. "Gaia would become a conscious, thinking, perceiving being functioning at a new evolutionary level with faculties quite literally beyond our imagination."

This is an inspiring concept, but we must not pass too quickly over Russell's qualification—*if* mankind were to evolve into a healthy, integrated social system. The two aspects of cooperative effort—building the society and developing the culture—are interdependent. A collective mind could only be realized in a peaceful and smoothly functioning society.

Although mankind possesses remarkable potential, there is no guarantee that he will continue to occupy his present position as the spearhead of the evolution of Form. The process advances tentatively by trial and error, two steps forward, one step back. The history of life on earth has shown many failed experiments. Lines that have seemed promising have often proved unfit to serve as the base for the next stage of development. The danger of the human condition lies in the fact that evolution has carried the development of the individual very rapidly to a high level of freedom, a freedom that has not yet been sufficiently tempered by the more slowly developing socializing influences and cultural conditioning. If the delicate balance between liberty and license, between personal initiative and cooperation with others, cannot be established soon, mankind will have failed to meet the evolutionary challenge. And some other form of life will probably evolve to fill the role.

In order to work effectively toward the creation of a suprahuman system, we must speed the evolution of the characteristics that promote synergistic relationships among all the single elements. In most advanced societies cultural conditioning has encouraged these socializing traits: consideration for others, altruism, selflessness, and compassion. In story and legend, in ballad and song, these traits are constantly held up as ideals of behavior. Laws, both written and unwritten, are founded on respect for individual rights and protection of equal opportunity for the development of inner potential.

Most of the great religions have recognized the importance of these unifying qualities. Christianity and Confucianism have enunciated nearly identical versions of the Golden Rule. Buddhists are taught to show "a boundless heart toward all human beings." The founders of these faiths appeared during a relatively brief period in human history, at a time when the formation of large urban civilizations was replacing the smaller, simpler units of tribal life. Codes of ethical behavior were needed so the new complex societies could function smoothly—so man could join with man in forming harmonious social systems with internal coordination that benefited all.

These influences, however, are on the wane today. As knowledge about the universe has grown, the acceptance of a code of behavior dictated by faith has been challenged. The role of revelation has been replaced by reason, and the study of nature has not found any scientific justification for these human values. In the last two centuries science has presented the universe first as a rigid and mindless mechanism and more recently as a meaningless collection of matter produced by random motion. In these world-schemes the ideals of personal worth, love, and charity find no rational support. They have been characterized as fairy tales with which mankind has consoled himself in a cold and indifferent universe.

A more optimistic interpretation suggests that we are living through a transitional period; if we can avoid the very real dangers that threaten mankind today, we can look forward to a time when a deeper understanding of the nature of the cosmos will lead to a recognition of our unique role in its evolution. A belief in our mission will replace the blind imperatives of faith and lead to the peaceful unification of all mankind.

We may hope that the form-building process will draw together the parts of this new organization, as it has forged the smaller units of mankind—a married couple, a tribe, a nation. A good marriage serves as a model for the kind of relationship that must be built into a suprahuman society. The synthesis of two lives which takes place within this context enhances the development of each participant while at the same time producing a more complete, more effective life force. As in a symbiotic relationship, each partner contributes to the welfare of the other; so the individual life is richer as a part than as a whole. The goal of a society built on this model may seem impossibly idealistic, but it may be the only *realistic* scenario for the future of mankind. As Buckminster Fuller once said: "The world is now too dangerous for anything less than Utopia."

The challenge is great because the future is open and is directly affected by everything we do. There are new worlds to be opened up, new faculties to be discovered, new powers to be revealed. As parts of a larger whole working together in a symbiotic relationship, we might indeed become the transient but essential elements in a new level of form. Life, mind, and culture will be integral parts of this greater organism. The new whole will, by definition, possess qualities which are not present in any of its parts alone. So the metaphors based on our present models—a cosmic mind, a universal culture, a nervous system for the planet—will, I suspect, be quite inadequate to describe the form that has not yet come into being. Our most imaginative projections will pale beside the reality that takes shape tomorrow.

CONCLUSION

I have discovered that I live in "creation's dawn."
The morning stars still sing together, and the
world, not yet half made, becomes more beautiful
every day.

JOHN MUIR, in
John of the Mountains

In the dark of the moon on a cloudless night, the star-studded universe is displayed in all its splendor. A curtain is lifted on this drama night after night, year after year, revealing the presence of mysterious worlds beyond our own, reminding us that we are part of something much greater than ourselves—stars more numerous than grains of sand on a beach, distances and time stretching out as far as the most powerful telescopes have penetrated and farther still. The mind cannot encompass such immensity. Almost unconsciously, we seek out familiar objects, little guideposts in the heavens so we will not be lost, awash in infinity.

On this warm, clear October night I have watched the slender crescent of the new moon intensify in the pale-pink sunset sky— and then the evening star. As darkness falls the hazy disk of the Milky Way begins to make a faint silver path almost directly overhead, and nearby Vega shines diamond-bright. Low in the east hangs the brilliant cluster of the Pleiades, their outlines diffused and softened by the luminous cloud in which they are immersed. None of the stars are perfectly steady and clear, of course; the images tremble slightly as their light passes through the thin halo of air that envelops the earth.

If we lived on a planet like Venus with an atmosphere dense enough to completely muffle the light of the stars and obscure the disc of the sun, we would have lived in ignorance of this vast shining universe. We would never have seen the flaming colors of the departing sun, nor watched a shower of shooting stars, nor followed the track of Orion across the sky. Imagine with what astonishment we would have viewed the cosmos for the first time if we had discovered how to rise above the planet's atmosphere in rockets and satellites. As the spacecraft rose through the dense atmosphere, an observer would see daylight slowly fade. The brightest stars would begin to shine palely through, inexplicable points of light in the haze. The horizon would gradually take shape, becoming more and more distinct, until it encircled the spacecraft. And straight above would be displayed the dome of heaven, velvety darkness set with countless tiny jewels of light.

The sense of wonder inspired by this strange and beautiful sight can hardly be imagined here on earth, where its impact has been dulled by long familiarity. We have been hypnotized and lulled by the simple repetitiveness of this drama, impressed night after night upon our consciousness. We are reassured by the daily rhythms that seem to imply stability and eternal return, although a study of the universe and the details of our own planet have shown us that nothing really repeats. The pattern that unwinds with time does not close back upon itself; it moves upward a little at each turning, like a coiled spring or the spiral of a mollusk shell (Plate 4). While the earth has made one revolution in its orbit around the sun, the solar system has traveled almost four and a half billion miles around the Milky Way, and the whole galaxy has moved four billion miles in the direction of the constellation Virgo. The very ground on which we stand has shifted on the planet's surface. The youngest mountains have increased in stature, while others are washed grain by grain into the sea. The Atlantic Ocean and the Red Sea have grown an inch or two wider. The passengers on Planet Earth have changed,

too. Species have become extinct and mutations have produced
new life forms.

The difficulty with perceiving real change and understanding
time as a dimension of creation is a matter of scale. The small
rhythmic patterns fall within a time span that comfortably
matches our own lives, while the great directional movements
take place on so much larger a scale that perception requires
a time-lapse view of cosmic history.

In this book of just two hundred pages we have traced the
life story of the universe from its birth to its present state. Now
we must compress the history even further—speed up the time-
lapse movie so we can see the growth and evolution of Form taking place before our eyes. Then we can appreciate the nature
and characteristics of the remarkable process that has made the
wondrously varied and beautiful universe from "almost nothing."

At the very beginning we see a flux of energy and matter suddenly start to expand, a plasma formless and disorganized but
intensely concentrated. Fractions of a second later elementary
particles of matter begin to emerge from the plasma. These particles come together, unite to make more complex units, and
build ever upward and outward as each organism extends its pattern in space-time. Protons are built from quarks, atoms from
protons and leptons, molecules and crystals from atoms. Each
higher stage of organization occurs by fusion and by internal rearrangement of the component parts to achieve greater stability
and longer life, so the levels are attained in steps, with no lasting intermediate forms. With increase of complexity the parts of
each whole are more diffusely arranged in space. The concentration of matter decreases; each organism occupies a greater volume.

As more advanced stages of organization are attained, more
options become available. The simplest inorganic molecules, for
example, have only one stable form; the complex organic molecules can be arranged in several configurations, producing different compounds; DNA molecules can assume an almost infinite
variety of forms.

At each stage of complexity new characteristics are realized
and the extension of Form is accelerated. Organisms act as patterns and catalysts, facilitating the emergence of new units. The

self-replicating molecule directs the formation of many others in its own likeness through countless generations. Or it may combine with another molecule, creating forms that never before existed. Matter flows in and out of these centers like miniature whirlpools in a turbulent stream. Extension in time—achieved by self-preservation and reproduction—creates the phenomenon of natural selection. As the building continues, the formative processes become less and less dependent on matter. In the most recent phase mind has revealed ways of extending Form even more efficiently in space and time. Embodied in symbols that can be infinitely reproduced, it is carried on waves of energy that travel with the speed of light.

As we stand here at this halfway station looking back at the history of this process, we can recognize certain outstanding features. It began slowly and is gathering momentum as it evolves. Generations of stars were required to synthesize the elements essential to life—a creative work that took perhaps ten billion years. Life is known to have existed three and a half billion years ago, and by one billion years ago multicellular organisms were present on the earth. Man-like species evolved from primitive stock sometime between ten and four million years ago, and *Homo sapiens* has been in existence for no more than a hundred thousand years. At each step the pace has quickened.

Spectacular as the advances have been, however, the process of transformation has moved forward in a tentative way, exploring a number of different solutions before the most effective ones are discovered. New options are constantly being tried, and even the most successful organisms are not in any sense final. They, too, must give way or become part of larger, more complete forms. Each new level of organization is only a way station on the road leading to the finished creation.

As we view the groping, exploratory nature of the process— the many unfavorable mutations, the tragic deformities—it is apparent that we are not witnessing the detailed accomplishment of a preconceived plan. "Nature is more and better than a plan in course of realization," Henri Bergson observed. "A plan is a term assigned to a labor: it closes the future whose form it indicates. Before the evolution of life, on the contrary, the portals of the future remain wide open. . . . This movement constitutes the unity of the organized world—a prolific unity, of an infinite richness, superior to any that the intellect could dream of, for the intellect is only one of its aspects or products."

The process of transformation has all the earmarks of a truly creative work in progress, bringing into being something that never existed before. As with all kinds of artistic creation, its beginnings were vague and formless; it has been taking shape ever since. Painters and sculptors, poets and musicians know that this tentative quality is characteristic of all creative processes. The artist works with the materials at hand, clarifying and crystallizing his original inspiration as he explores the hidden potentials of the medium. The poet searches for the words and rhythms and metaphors that evoke the images which will express his vision. "A picture," said Pablo Picasso, "is not thought out and determined beforehand, rather while it is being made it follows the mobility of thought." Roger Sessions described the composition of a piece of music in these terms:

> After inspiration and conception comes execution. The process of execution is first of all that of listening inwardly to the music as it shapes itself; of allowing the music to grow; of following both inspiration and conception wherever they may lead. A phrase, a motif, a rhythm, even a chord, may contain within itself, in the composer's imagination, the energy which produces movement. It will lead the composer on, through the force of its own momentum or tension, to other phrases, other motifs, other chords. . . . It seems to me . . . that art is a function, an activity of the inner nature—that the artist's effort is, using the raw and undisciplined materials with which his inner nature provides him, to endow them with a meaning which they do not of themselves possess— to transcend them by giving them artistic form.

A creative work cannot be produced by reason alone, because it is not a simple logical progression from A to B to C. Its progress has many turnings, hours of despair and moments of joy, illuminated by flashes of imagination and whimsy. These traits are clearly exemplified in the varied, fanciful, and seemingly capricious history of life on earth. "For naïve and unbiased contemplation nature does not look like a calculating merchant," said Ludwig von Bertalanffy;

> rather she looks like a whimsical artist, creative out of an exuberant fantasy. . . . she produces a wealth of colour, form, and other creations, which, as far as we can see, is completely useless. Consider, for example, the exquisite artistry of butterflies' wings, which has nothing to do with function, and cannot even be appreciated by their bearers with their imperfect eyes. This productivity and joy of creation seems to be expressed in the 'horizontal' multiplicity of forms on the same level of organization as well as in the 'vertical' progress of organization, which can, but need not necessarily, be considered as 'useful'. . . . Evolution appears to be more than the mere product of chance governed by profit. It seems a cornucopia of *évolution créatrice.*

One miniature masterpiece after another has spilled out of this cornucopia of creation: the clean mathematical perfection of the nautilus shell (Plate 15); the delicate design of the snowflake in almost infinite variety (Plate 5); the light and supple feather that carries the seagull soaring on graceful wings far above the planet's surface (Plate 7). In so many ways exuberance and joy are expressed in the shape and texture of little things. Sometimes the artistry is displayed for every eye to see, and sometimes it is hidden deep inside. Consider, for example, the seed of columbine—a tiny speck of matter appearing as inert and insignificant as a mote of dust, yet it holds within itself infinite potential: flower within seed, seed within flower, multiplying endlessly into the future. Summer after summer this seed transforms the upland meadows of our western mountains with floods of pale-blue blossoms, rippling seas of flowers. This is just one of many little miracles that are revealed to us each day, events so familiar that our sense of wonder is dulled, as with the view of the stars at night.

But year by year large-scale changes are taking place, too, carrying these miniature creations like diatoms borne on a rising tide. The process of transformation is not only accelerating but evolving as time goes by. Until the advent of the human brain, all change occurred through trial and error. Now the possibility of foresight has entered the cosmos. Although imperfectly developed (and frequently as inaccurate as the weather forecast for the next few days) this new power is beginning to speed the process of evolution, which is no longer totally dependent on the slow mechanism of natural selection. And the potential for further improvement is enormously great.

The evolution of ideas has added a new dimension to the creative process, supplementing but not displacing the evolution of material things. As mind becomes less and less dependent on matter a collective mind is taking shape—thoughts are reverberating around the earth and beginning to reach out into the universe.

Although mankind appears to be just a minute local phenomenon in a cosmos so vast that its size humbles the imagination, size alone is not a measure of importance. We have seen that the transformation process takes place by building from tiny individual centers. The whole is immanent in all the parts, no matter how small.

An exponential process magnifies in a spectacular way even very little beginnings. As the extension of the creative force of life and mind doubles and doubles again it will rapidly encompass ever widening spheres. We can imagine that the next stages of evolution will be more beautiful as they approach nearer to the ultimate expression of Form. Although the general direction of the transformation process can be perceived, we know that it is always moving into unexplored territory. Each new level of organization reveals qualities that cannot be anticipated until that stage is reached. (It would not have been possible, for example, to predict the creation of mind by studying the individual human cell.) Higher levels of organisms may be taking shape—not just in time but also in space, far out in the larger dimensions of the universe.

For many eons human beings observed the sky at night and

puzzled over the meaning of those myriad, moving points of light. They noted certain regularities in the movements and devised theories to account for them—theories which in general proved to be inconsistent with reality. But understanding of the universe has been growing rapidly in recent years; mankind has begun to learn how pieces of this puzzle fit together into larger systems. The earth is an integral part of the solar system, which is part of the spinning disk of stars whose inside edge looks like a hazy path across the heavens. And this Milky Way is one of nineteen galaxies in the Local Group which is part of a much larger cluster containing thousands of galaxies. By computer-assisted plotting we have found that the shape of this Local Supercluster resembles a giant many-petaled flower. Beyond this assembly, our most powerful telescopes reveal the presence of at least one billion galaxies distributed throughout the observable universe. Cosmologists are beginning to recognize that these are not just separate unrelated whirls of matter moving in lonely paths through space. They tend to cluster; they frequently collide and unite. Perhaps they are related to each other in more subtle ways that we have not yet perceived. Could they be single cells in a much greater organism?

These are revolutionary thoughts; but history has shown us that the truth always turns out to be more wonderful than anything we have imagined, or in fact that we *can* imagine. As Fred Hoyle so aptly expressed it: "If there is one important result that comes out of our inquiry into the nature of the Universe it is this: when by patient inquiry we learn the answer to any problem, we always find, both as a whole and in detail, that the answer thus revealed is finer in concept and design than anything we could ever have arrived at by a random guess."

The end product of the creative process cannot be known until the work has been completed. Like a symphony or a poem half-composed, the last movements have not yet been orchestrated, the last stanzas not yet set in verse. As the poet Brewster Ghiselin said, "I did not understand all these things or grasp their significance until I wrote the final lines."

The universe is unfinished, not just in the limited sense of an incompletely realized plan but in the much deeper sense of a

The whole is immanent in all the parts, no matter how small.

Plate 17
The Andromeda Galaxy (Lick Observatory Photograph).

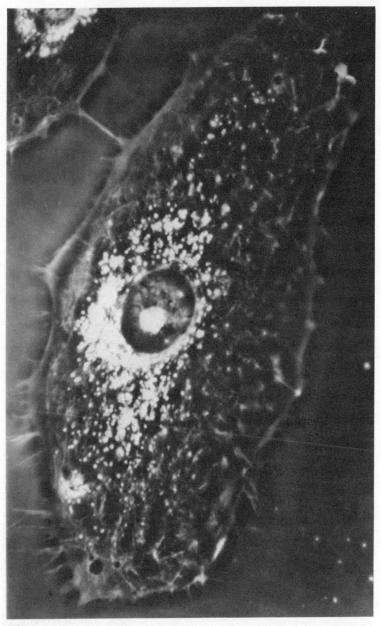

Plate 18
A Human Cell (Electron-microscope photograph by George O. Gey).

creation that is a living reality of the present. A masterpiece of artistic unity and integrated Form, infused with meaning, is taking shape as time goes by. But its ultimate nature cannot be visualized, its total significance grasped, until the final lines are written.

NOTES

INTRODUCTION

page 11 *Shaw quotation:* from George Bernard Shaw, *Back to Methuselah*, Preface.

pages 11–14 *Becker quotation:* from Carl L. Becker, *The Heavenly City of the Eighteenth Century Philosophers,* cited in *The Mystery of Matter,* edited by Louise B. Young, p. 466.

page 14 *Jeans quotation:* from James Jeans, *The Mysterious Universe,* p. 159.

page 14 *Smith quotation:* from Huston Smith, "The Revolution in Western Thought," cited in *The Mystery of Matter,* edited by Louise B. Young, p. 467.

page 15 *Weinberg quotation:* from Steven Weinberg, *The First Three Minutes,* pp. 154–155.

page 16 *Spender quotation:* as cited by Brewster Ghiselin, *The Creative Process,* Introduction, p. 14.

page 16 *Ghiselin quotation:* from Brewster Ghiselin, *The Creative Process,* p. 131.

CHAPTER 1

page 19 *Yeats quotation:* cited by J. T. Fraser, *The Voices of Time,* p. 58.

page 21 *Cyclical phenomena and the Brahma day:* see Joseph Needham, "Time and Knowledge in China and the West," *The Voices of Time,* pp. 129–130.

page 21 *Philosophy of Heracleitus and quotation:* see John Burnet, *Early Greek Philosophy,* quotation numbered fr. 41. See also Will Durant, *The Story of Philosophy,* pp. 73–74.

page 22 *Concept of the Great Year:* see J. L. Russell, "Time in Christian Thought," *The Voices of Time,* p. 68.

pages 22–23 *Every event infinitely repeated:* ibid., pp. 68–69.

page 23   *Needham quotation:* from Joseph Needham, "Time and Knowledge in China and the West," *The Voices of Time,* p. 129.

page 24   *Christian concept of time:* ibid., p. 130.

page 24   *Second Coming:* Mark 13:35 (Authorized Version).

pages 24–25   *Eighty spheres required:* see Herbert Butterfield, *The Origin of Modern Science: 1300–1800,* p. 40.

page 26   *Theory of seed containing potential of form:* see Martin J. S. Rudwick, *The Meaning of Fossils,* p. 84.

page 26   *Hooke's theory of extinction:* ibid., pp. 64–68.

page 27   *Hutton quotation:* from James Hutton, *Theory of the Earth,* p. 304.

page 27   *Casanova quotation:* Casanova de Seingalt, *Mémoires,* cited by Edward White, "The Chinese Landscape," *Horizon,* Autumn 1975, p. 97.

page 28   *Hutton quotation:* from James Hutton, *Theory of the Earth,* p. 304.

page 28   *Influence of religion on Lamarck's theory:* see Martin J. S. Rudwick, *The Meaning of Fossils,* pp. 118–119.

page 29   *Lyell's theories of geology:* see Charles Lyell, *Principles of Geology.*

page 29   *Rudwick quotation:* from Martin J. S. Rudwick, *The Meaning of Fossils,* p. 179.

page 30   *Bergson quotation:* from Henri Bergson, *Creative Evolution,* pp. 38–39.

page 31   *Trefil quotation:* from James Trefil, "How the Universe Will End," p. 83.

page 31   *Khayyam quotation:* from Omar Khayyam, *The Rubáiyát,* LIII.

page 34   *Darwin quotation:* from Charles Darwin, *Origin of Species,* last paragraph.

pages 34–35   *Huxley quotation:* from Julian Huxley, *Evolution in Action,* pp. 9–10.

page 35   *Teilhard's philosophy:* see Pierre Teilhard de Chardin, *The Phenomenon of Man.*

pages 35–36   *Teilhard quotation:* ibid., p. 218.

CHAPTER 2

pages 37–38   *Cosmologists expect to find simplicity:* see James Trefil, "Closing in on Creation," p. 42.

page 38   *Guth quotation:* cited by Rick Gore, "The Once and Future Universe," p. 741.

page 39 *Disintegration rate of proton:* see James Trefil, "Closing in on Creation," p. 41.

pages 41–42 *Whitehead quotation:* from Alfred North Whitehead, *Science and the Modern World,* p. 105.

page 42 *Odds of achieving balance of expansion and contraction:* Michael Turner as quoted by Rick Gore, "The Once and Future Universe," pp. 741, 744.

page 51 *Rate of crystal growth:* see Paul E. Desautels, *The Mineral Kingdom,* pp. 48–49: "sixteen trillion atoms an hour."

page 51 *Age of obsidian:* see Peter Francis, *Volcanoes,* p. 156.

page 52 *Molecules in space:* see Richard McCray, "Molecules between the Stars," p. 74.

page 52 *Species of interstellar molecules:* see Richard H. Gammon, "Chemistry of Interstellar Space," p. 30.

page 53 *Time estimates for destruction and creation of molecules in space:* see Richard McCray, "Molecules between the Stars," p. 75.

page 53 *Symbiotic relationship between molecules and stars:* suggested by Richard H. Gammon, "Chemistry of Interstellar Space," p. 30.

page 54 *Star formation a self-regulating process:* suggested by Robert D. Gehrz *et al.,* "The Formation of Stellar Systems from Interstellar Molecular Clouds," pp. 824–825.

page 54 *Star formation in the Great Nebula:* see Richard McCray, "Molecules between the Stars," p. 73.

page 54 *Star formation in Tarantula Nebula:* see Rick Gore, "The Once and Future Universe," p. 727.

pages 55–56 *X-ray and gamma-ray universe:* ibid., pp. 728–730.

page 56 *Jeans quotation:* from James Jeans, *The Mysterious Universe,* p. 6.

page 56 *Barrow quotation:* cited by Rick Gore, "The Once and Future Universe," p. 745.

CHAPTER 3

page 57 *Quotation:* from Loren Eiseley, *The Immense Journey,* p. 209.

page 59 *Molecules in ancient rocks:* see "Evidence of Life Found in Oldest Known Rocks," *Chemical and Engineering News,* September 17, 1979, pp. 22–23.

pages 60–61 *Murchison meteorite:* see "News Notes," *Sky and Telescope,* December 1969, p. 388.

page 61 *Meteorites found in Antarctica:* see Ursula B. Marvin, "Meteorites on Ice," p. 6.

page 62   *Claims of fossil algae in meteorites:* George Claus and Bartholomew Nagy were the biochemists. See Brian Mason, "Organic Matter from Space," p. 44.

page 62   *Cell-like contaminants in meteorites:* described by Lawrence Grossman, University of Chicago, personal communication, 1980.

page 63   *Research at University of Maryland:* reported by Ronald Kotulak, "Extraterrestrial Life?" *Chicago Tribune,* October 30, 1983, Section 6, pp. 1–2. Personal communication with Mitchell Hovish, University of Maryland, March 20, 1985.

page 64   *Eagerness to associate:* see Ronald Kotulak, "Extraterrestrial Life?" p. 2.

page 65   *Clay as a template:* see E. T. Degens *et al.,* "Template Catalysis: Asymmetric Polymerization of Amino Acids on Clay," p. 492. See also M. Paecht-Horowitz *et al.,* "Prebiotic Synthesis of Polypeptides . . . ," pp. 636–639.

pages 67–68   *Coacervates:* see Richard E. Dickerson, "Chemical Evolution and the Origin of Life," pp. 83–86.

page 69   *Ponnamperuma quotation:* cited by Ronald Kotulak, "Extraterrestrial Life?" p. 2.

page 74   *Eiseley quotation:* from Loren Eiseley, *The Immense Journey,* pp. 26–27, 210.

CHAPTER 4

page 77   *Warrawoona Formation:* personal communication with John Sepkowski, University of Chicago, 1983.

pages 77–78   *Theory of cells uniting to form larger wholes:* see Lynn Margullis, *Symbiosis in Cell Evolution,* pp. 1–14.

page 79   *Use of oxygen by living organisms:* see Preston Cloud and Aharon Gibor, "The Oxygen Cycle."

page 81   *Conjugation:* see H. G. Wells *et al., The Science of Life,* pp. 443–444.

page 83   *Long chain colonies as first multicellular plants:* ibid., pp. 662–663.

pages 88–89   *Nitrogenase and the fixation of nitrogen:* see C. C. Delwiche, "The Nitrogen Cycle," p. 142. See also Lynn Margulis, *Symbiosis in Cell Evolution,* pp. 172, 179.

pages 90–91   *Huxley quotation:* from Julian Huxley, *Evolution: The Modern Synthesis,* cited in *The Evolution of Man,* edited by Louise B. Young, pp. 170–171.

page 91   *Dobzhansky quotation:* from Theodosius Dobzhansky, *The*

Biology of Ultimate Concern, cited in *The Evolution of Man,* edited by Louise B. Young, pp. 147–149.

pages 92–93 *Recapitulation:* see H. G. Wells *et al., The Science of Life,* pp. 152–154, 366–373.

page 93 *Number of atoms and genes in chromosomes:* Ibid., pp. 494, 498, 598.

pages 93–94 *Comparison of human and great ape chromosomes:* see Jorge J. Yunis and Om Prakash, "The Origin of Man: A Chromosomal Pictorial Legacy," pp. 1525–1530.

CHAPTER 5

pages 98–99 *Life cycle of cellular mold:* see Maurice Sussman, *Growth and Development,* pp. 25–34.

pages 99–100 *Experiments with sponges:* see Edmund W. Sinnott, *The Biology of the Spirit,* pp. 32–36.

page 101 *Life cycle of Hylaeus:* see Edwin Way Teale, *The Golden Throng,* p. 26.

page 101 *Life cycle of mason bees:* ibid., p. 29.

pages 101–102 *Life cycle of Halictidae:* ibid., pp. 29–30.

page 102 *Life cycle of bumblebees:* ibid., pp. 30–32.

pages 102–103 *Life cycle of honeybees:* ibid., pp. 36–43.

pages 103–104 *Lichens:* see Sylvia Duran Sharnoff, "Lowly Lichens Offer Beauty—and Food, Drugs, and Perfume," pp. 135–143.

page 104 *Reproductive processes of lichen:* see H. G. Wells *et al., The Science of Life,* pp. 292–295.

pages 108–109 *Symbiotic relationship of zooxanthellae and coral polyps:* see R. E. Johannes, "Life and Death of the Reef," p. 45.

page 109 *Life cycle of coral polyp:* see Robert F. Sisson, "Life Cycle of a Coral," p. 780.

pages 110–111 *Diver's description of cleaner shrimp:* quoted by René Dubos, *The Torch of Life,* pp. 68–69.

pages 111–112 *The coral reef as an organism:* see Richard Chesher, *Living Corals,* pp. 28–30.

page 113 *Thomas quotation:* from Lewis Thomas, *The Lives of a Cell,* p. 145.

pages 113–114 *Gaia theory:* from J. E. Lovelock, *Gaia: A New Look at Life on Earth.*

page 114 *Margulis quotation:* cited by J. E. Lovelock, ibid., p. 128.

pages 115–116 *Atkins quotation:* from P. W. Atkins, *The Creation,* p. 123.

page 116   *Muir quotation:* from John Muir, *Gentle Wilderness: The Sierra Nevada*, p. 146.

CHAPTER 6

pages 117–118   *Kelvin's work:* see William Garnett and Hugh Munro Ross, "Kelvin," pp. 722–724.

page 118   *Age of the earth:* see William Thomson (Lord Kelvin), "Secular Cooling of the Earth."

page 125   *Distribution of gold:* see William Elmer Caldwell, "Gold," pp. 479–481.

page 130   *Schrödinger quotation:* from Edwin Schrödinger, *What Is Life? & Other Scientific Essays*, cited in *The Mystery of Matter*, edited by Louise B. Young, p. 435.

page 131   *Example of increased life expectancy of molecules:* ibid., p. 439.

page 131   *Peattie quotation:* from Donald Culross Peattie, *An Almanac for Moderns*, p. 12.

pages 132–133   *Krutch quotation:* from Joseph Wood Krutch, *The Great Chain of Life*, pp. 80–81.

page 136   *Teilhard quotation:* from Pierre Teilhard de Chardin, *The Phenomenon of Man*, p. 51.

CHAPTER 7

pages 141–142   *Huxley quotation:* from Julian Huxley, *Evolution in Action*, pp. 40–42.

page 144   *Krutch quotation:* from Joseph Wood Krutch, *The Great Chain of Life*, pp. 167–168.

pages 145–146   *Bartalanffy quotation:* from Ludwig von Bertalanffy, *Problems of Life*, pp. 95–96, 103.

page 147   *Bronowski quotation:* from Jacob Bronowski, *The Commonsense of Science*, p. 88.

CHAPTER 8

page 151   *Dostoevsky quotation:* from Feodor Dostoevsky, *The Brothers Karamazov*, Book V, Chapter V.

pages 151–152   *Chromosomes of man and great apes:* see Jorge J. Yunis and Om Prakash, "The Origin of Man: A Chromosomal Pictorial Legacy," pp. 1525–1529.

page 152   *Differences in body proteins:* see S. L. Washburn and Ruth Moore, *Ape into Man*, pp. 11–22.

pages 152–153   *Early evolution of man:* ibid., p. 166.

page 153 *Finch in Galápagos:* see the photo of a tool-using finch by Eibl-Eibesfeldt in *The Evolution of Man,* edited by Louise B. Young, p. 29.

page 153 *Use of tools by chimps:* see Jane van Lawick-Goodall, *My Friends the Wild Chimpanzees,* pp. 31–36, 48–50, 76–77.

page 153 *Hominids making stone tools:* see J. W. K. Harris, "Delving into Early Human Behavior: Life on the Koobi Fora Savanna," p. 11.

page 154 *Fires laid 500,000 years ago:* see S. L. Washburn and Ruth Moore, *Ape into Man,* p. 185. See also Carl Sagan, *The Dragons of Eden,* p. 88.

page 155 *Growth of brain size after birth:* see Loren Eiseley, *The Immense Journey,* p. 109.

page 155 *Immaturity in human babies:* see Julian Huxley, *Man in the Modern World,* cited in *The Evolution of Man,* edited by Louise B. Young, p. 244.

page 155 *Biblical quotation:* Genesis 3:16 (Authorized Version).

page 156 *Community found in Afar depression:* see J. W. K. Harris, "Delving into Early Human Behavior," p. 10.

page 156 *Wild dogs in Africa:* see Félix Rodríguez de la Fuente, *World of Wildlife: I. Africa: Hunters and Hunted of the Savannah,* pp. 241–263.

pages 156–157 *Great ape communities:* see S. L. Washburn and Ruth Moore, *Ape into Man,* pp. 133–134.

page 157 *Sharing of food in ape communities:* ibid., p. 134. See also Jane van Lawick-Goodall, *My Friends the Wild Chimpanzees,* p. 93.

pages 157–160 *Benedict quotation:* from Ruth Benedict, *Patterns of Culture,* pp. 46–47.

page 160 *Dobzhansky quotation:* from Theodosius Dobzhansky, *The Biological Basis of Human Freedom,* p. 43.

pages 161–162 *Experiments on snails:* see Jeffrey L. Fox, "Debate on Learning Theory Is Shifting," p. 1219.

pages 162–163 *David Graybeard incident:* see Jane van Lawick-Goodall, *In the Shadow of Man,* pp. 66–68.

page 163 *Goodall quotation:* from Jane van Lawick-Goodall, *My Friends the Wild Chimpanzees,* pp. 137–138.

pages 163–164 *Ape and human linguistic abilities:* see S. L. Washburn and Ruth Moore, *Ape into Man,* p. 176.

page 164 *Language and Homo sapiens:* ibid., p. 178.

page 165 *Eiseley quotation:* from Loren Eiseley, *The Firmament of Time,* pp. 112–114.

CHAPTER 9

page 168   *Myths and legends about rainbows:* see Carl Boyer, "Art, Myth, and Magic."

page 170   *Thinking about the world in a Greek way:* see John Burnet, *Early Greek Philosophy,* Preface to Fourth Edition.

page 170   *Philosophy of Thales, Democritus, and Pythagoras:* see Benjamin Farrington, *Greek Science;* Giorgio de Santillana, *The Origins of Scientific Thought.*

pages 170–171   *Whitehead quotation:* from Alfred North Whitehead, *Science and the Modern World,* pp. 4–7.

pages 171–172   *Newton quotation:* from Isaac Newton, "Dispersion of Light," in *Moments of Discovery,* pp. 393–400.

pages 172–173   *Schrödinger quotation:* from Erwin Schrödinger, *What Is Life? & Other Scientific Essays,* cited in *The Mystery of Matter,* edited by Louise B. Young, p. 453.

pages 176–177   *Fractals:* see Jeanne McDermott, "Geometric Forms Known as Fractals Find Sense in Chaos." See also Dietrick Thomsen, "Making Music—Fractally," and "A Place in the Sun for Fractals."

page 177   *Popper's example:* cited by Jacob Bronowski, *Science and Human Values,* pp. 24–25.

page 178   *Ockham's razor: Entia non sunt multiplicanda praeter necessitatem,* William of Ockham, *Commentarium in Libros IV Sententiarum Petri Lombardi* (1318–23).

page 178   *Eddington quotation:* cited by Tobias Dantzig, *Number, the Language of Science,* p. 230.

pages 178–179   *Kepler's work:* see Arthur Koestler, *The Sleepwalkers,* pp. 312–334.

page 179   *Koestler quotation:* from Arthur Koestler, *The Sleepwalkers,* cited in *Exploring the Universe,* edited by Louise B. Young, p. 240.

page 181   *Darwin's response to Wallace's paper:* see Loren Eiseley, *The Immense Journey,* p. 86.

page 181   *Sinnott quotation:* from Edmund Sinnott, *Matter, Mind, and Man,* p. 130.

page 182   *Lines of poetry:* from Carl Sandburg, "Fog"; Robinson Jeffers, "Night"; Amy Lowell, "Night Clouds."

pages 182–183   *MacLeish quotation:* from Archibald MacLeish, "The Great American Frustration," p. 14.

page 183   *Beckett quotation:* from Samuel Beckett, *The Collected Works of Samuel Beckett: Waiting for Godot,* p. 22.

pages 183–184   *Russell quotation:* from Bertrand Russell, "A Free

Man's Worship," cited in *Exploring the Universe,* edited by Louise B. Young, pp. 517–518.

page 185   *Sinnott quotation:* from Edmund Sinnott, *Matter, Mind, and Man,* p. 185.

CHAPTER 10

pages 186–187   *Galton quotation:* from Francis Galton, *Inquiries into Human Faculty and Its Development,* cited in *The Evolution of Man,* edited by Louise B. Young, p. 368.

page 188   *Wells quotation:* from H. G. Wells *et al., The Science of Life,* pp. 295–299.

page 189   *Spencer quotation:* from Herbert Spencer, *Principles of Sociology,* cited in *The Evolution of Man,* edited by Louise B. Young, pp. 361–362.

pages 189–190   *Dubos quotation:* from René Dubos, *The Torch of Life,* p. 102.

pages 191–192   *Wells quotation:* from H. G. Wells *et al., The Science of Life,* p. 1475.

page 192   *Teilhard quotation:* from Pierre Teilhard de Chardin, *The Phenomenon of Man,* p. 261.

page 193   *Teilhard quotation:* ibid., pp. 251–252.

pages 194–195   *Thomas quotation:* from Lewis Thomas, *The Medusa and the Snail,* p. 15.

page 195   *Russell quotation:* from Peter Russell, *The Global Brain,* p. 231.

page 197   *Fuller quotation:* cited by Peter Russell, *The Global Brain,* p. 199.

CONCLUSION

page 199   *Movements of the earth and the galaxy:* see M. Mitchell Waldrop, "A Flower in Virgo," p. 953.

pages 201–202   *Bergson quotation:* from Henri Bergson, *Creative Evolution,* pp. 103–105.

page 202   *Picasso quotation:* from Christian Zervos, "Conversations with Picasso."

page 202   *Sessions quotations:* from Roger Sessions, "The Composer and His Message."

page 203   *Bertalanffy quotation:* from Ludwig von Bertalanffy, *Problems of Life,* pp. 106–108.

page 205   *Shape of Local Supercluster:* see M. Mitchell Waldrop, "A Flower in Virgo," p. 953.

page 205   *Hoyle quotation:* from Fred Hoyle, *The Nature of the Universe,* p. 141.

page 205   *Ghiselin quotation:* from Brewster Ghiselin, *The Creative Process, A Symposium,* p. 133.

BIBLIOGRAPHY

Atkins, P. W. *The Creation*. Oxford and San Francisco: W. H. Freeman & Company, 1981.

Becker, Carl. *The Heavenly City of the Eighteenth Century Philosophers*. New Haven: Yale University Press, 1932.

Beckett, Samuel. *The Collected Works of Samuel Beckett: Waiting for Godot*. New York: Grove Press, Inc., 1954.

Benedict, Ruth. *Patterns of Culture*. Boston: Houghton Mifflin Company, 1934. Sentry Edition, 1959.

Bergson, Henri. *Creative Evolution*. Translated from the French by Arthur Mitchell. New York: Henry Holt and Company, 1913.

Bertalanffy, Ludwig von. *Problems of Life: An Evaluation of Modern Biological and Scientific Thought*. New York: Harper & Brothers, 1952.

Boyer, Carl. "Art, Myth, and Magic." *The Rainbow Book*. The Fine Arts Museums of San Francisco in association with Shambhala. Berkeley and London: Fine Arts Museums of San Francisco, 1975.

Bronowski, Jacob. *The Commonsense of Science*. Cambridge: Harvard University Press, 1953.

———. *Science and Human Values*. New York: Harper & Brothers, 1956.

Burnet, John. *Early Greek Philosophy*. London: A. & C. Black, 1930.

Butterfield, Herbert. *The Origins of Modern Science: 1300–1800*. New York: The Macmillan Company, 1957.

Caldwell, William Elmer. "Gold." In *Encyclopedia Brittanica* (1956), vol. 10, pp. 479–481.

Chesher, Richard, and Douglas Faulkner (photographer). *Living Corals*. New York: Clarkson N. Potter, Inc., 1979.

Cloud, Preston, and Aharon Gibor. "The Oxygen Cycle." *Scientific American*, vol. 223, no. 3 (September 1970), pp. 110–123.

219

Dantzig, Tobias. *Number, the Language of Science.* New York: The Macmillan Company, 1944.

Darwin, Charles. *Origin of Species.* London: J. Murray, 1st edition 1859, 6th edition 1872.

Degans, E. T., *et al.* "Template Catalysis: Asymmetric Polymerization of Amino Acids on Clay." *Nature,* vol. 227 (August 1, 1970), p. 492.

Delwiche, C. C. "The Nitrogen Cycle." *Scientific American,* vol. 223, no. 3 (September 1970), pp. 136–147.

Desautels, Paul E. *The Mineral Kingdom.* New York: Madison Square Press, 1968.

Dickerson, Richard E. "Chemical Evolution and the Origin of Life." *Scientific American,* vol. 239, no. 3 (September 1978), pp. 83–86.

Dobzhansky, Theodosius. *The Biological Basis of Human Freedom.* New York: Columbia University Press, 1956.

———. *The Biology of Ultimate Concern.* Perspectives in Humanism Series, edited by Ruth Nanda Anshen. New York: The New American Library, 1967.

Dostoevsky, Feodor. *The Brothers Karamazov.* Translated by Constance Garnett. New York: The Modern Library, 1950.

Dubos, René. *The Torch of Life.* The Credo Series, planned and edited by Ruth Nanda Anshen. New York: Simon & Schuster, 1962.

Durant, Will. *The Story of Philosophy.* Garden City, NY: Garden City Publishing Co., 1926.

Eiseley, Loren. *The Immense Journey.* New York: Random House, 1946.

———. *The Firmament of Time.* New York: Atheneum, 1962.

"Evidence of Life Found in Oldest Known Rocks." *Chemical and Engineering News* (September 17, 1979), pp. 22–23.

Farrington, Benjamin. *Greek Science.* Harmondsworth, England and Baltimore, Md.: Penguin Books, Ltd., 1949.

Fox, Jeffrey L. "Debate on Learning Theory Is Shifting." *Science,* vol. 222, no. 4629 (December 16, 1983), pp. 1219–1222.

Francis, Peter. *Volcanoes.* London: Penguin Books, Ltd., 1976.

Fraser, J. T., editor. *The Voices of Time.* New York: George Braziller, 1966.

Gammon, Richard H. "Chemistry of Interstellar Space." *Chemical and Engineering News* (October 2, 1978), pp. 21–33.

Garnett, William, and Hugh Munro Ross. "Kelvin." In *Encyclopaedia Britannica* (1911), vol. 15, pp. 722–724.

Gehrz, Robert D., David C. Black, and Philip M. Solomon. "The Formation of Stellar Systems from Interstellar Molecular Clouds." *Science,* vol. 224, no. 4651 (May 25, 1984), pp. 823–830.

Ghiselin, Brewster, editor. *The Creative Process.* Berkeley: University of California Press, 1952.

Goodall, *see* Lawick-Goodall.

Gore, Rick. "The Once and Future Universe." *National Geographic,* vol. 163, no. 6 (June 1983), pp. 704–749.

Hanna, John Muir, and Ralph Eugene Wolfe. *John of the Mountains.* Boston: Houghton Mifflin, 1938.

Harris, J. W. K. "Delving into Early Human Behavior: Life on the Koobi Fora Savanna." *The L. S. B. Leakey Foundation News.* Pasadena, CA: The Leakey Foundation, Spring 1981.

Hoyle, Fred. *The Nature of the Universe.* New York: Harper & Row Publishers, Inc., 1950; revised edition, 1960.

Hutton, James. *Theory of the Earth.* Edinburgh: *Transactions of the Royal Society of Edinburgh,* 1788.

Huxley, Aldous. *Science, Liberty, and Peace.* New York and London: Harper & Brothers Publishers, 1946.

Huxley, Julian. *Evolution in Action.* New York: Harper & Brothers, 1953; New American Library edition, 1953.

———. *Evolution: the Modern Synthesis.* New York: Harper & Row Publishers, Inc., 1942.

———. *Man in the Modern World.* New York: Harper & Row Publishers, Inc., 1939.

Ionesco, Eugène. *Rhinocerus and Other Plays.* Translated by Derek Prouse. New York: Grove Press, Inc., 1960.

Jastrow, Robert. *Until the Sun Dies.* New York: W. W. Norton and Company, Inc., 1977.

Jeans, James. *The Mysterious Universe.* New York: The Macmillan Company, 1930.

Johannes, R. E. "Life and Death of the Reef." *Audubon* (September 1976), pp. 38–55.

Johanson, Donald. "Lucy." *University of Chicago Magazine.* Chicago: University of Chicago (Spring 1981), pp. 4–9.

Jung, Carl. *Memories, Dreams, Reflections.* Recorded and edited by

Aniela Jaffe, translated from the German by Richard and
Clara Winston. New York: Random House, 1965.

Kafka, Franz. *The Penal Colony.* Translated by Willa and Edwin Muir.
New York: Schocken Books, 1948.

Khayyam, Omar. *The Rubáiyát of Omar Khayyam.* Translated by Edward Fitzgerald. New York: The Heritage Press, 1946.

Koestler, Arthur. *The Sleepwalkers.* New York: The Macmillan Company, 1959.

Kotulak, Ronald. "Extraterrestrial Life?" *Chicago Tribune,* October 30,
1983, Section 6, pp. 1–2.

Krutch, Joseph Wood. *The Great Chain of Life.* Boston: Houghton
Mifflin Company, 1957; New York: Pyramid Books, 1966.

Lawick-Goodall, Jane van. *My Friends the Wild Chimpanzees.* Washington, D.C.: National Geographic Society, 1967.

————. *In the Shadow of Man.* Boston: Houghton Mifflin Company,
1971.

Lillie, Ralph. *General Biology and Philosophy of Organism.* Chicago:
University of Chicago Press, 1945.

Lovelock, J. E. *Gaia: A New Look at Life on Earth.* Oxford: Oxford
University Press, 1979.

Lyell, Charles. *Principles of Geology.* London: J. Murray, 1830–33.

MacLeish, Archibald. "The Great American Frustration," *Saturday Review* (July 13, 1968), pp. 13–16.

McCray, Richard. "Molecules between the Stars," *Natural History*
(December 1974), pp. 72–77.

McDermott, Jeanne. "Geometric Forms Known as Fractals Find Sense
in Chaos." *Smithsonian* (December 1983), pp. 110–117.

Margulis, Lynn. *Symbiosis in Cell Evolution.* San Francisco: W. H.
Freeman and Company, 1981.

Marvin, Ursula B. "Meteorites on Ice." *The Planetary Report* (March/
April 1984), pp. 5–7.

Mason, Brian. "Organic Matter from Space," *Scientific American* (March
1963), pp. 44f.

Muir, John. *Gentle Wilderness: The Sierra Nevada.* Edited by David
Brower. New York: Ballantine Books, 1967.

Needham, Joseph. "Time and Knowledge in China and the West." In
The Voices of Time, edited by J. T. Fraser. New York:
George Braziller, 1966, pp. 92–135.

"News Notes." *Sky and Telescope* (December 1969), p. 388.

Newton, Isaac. "Dispersion of Light." In *Moments of Discovery,* edited
by George Schwartz and Philip W. Bishop. New York:

Basic Books, Inc., 1958, vol. 1, pp. 393–400. Originally appeared in *Philosophical Transactions of the Royal Society of London,* vol. 1, 1672.

Paecht-Horowitz, M., J. Berger, and A. Katchalsky. "Prebiotic Synthesis of Polypeptides by Heterogenous Polycondensation of Amino-acid Adenylates." *Nature,* vol. 228 (November 14, 1970), pp. 636–639.

Peattie, Donald Culross. *An Almanac for Moderns.* New York: G. P. Putnam's Sons, 1935.

Porter, Roy. *The Making of Geology: Earth Science in Britain, 1660–1815.* Cambridge: Cambridge University Press, 1977.

Rodríguez de la Fuenta, Féliz. *World of Wildlife.* Translated by John Gilbert. London: Orbis Publishing Limited, 1970.

Roessler, Carl. *The Underwater Wilderness: Life Around the Great Reefs.* New York: Chanticleer Press, 1977.

Rudwick, Martin J. S. *The Meaning of Fossils.* New York: Science History Publications, 2nd edition, 1976.

Russell, Bertrand. "A Free Man's Worship." In *Mysticism and Logic.* London: George Allen & Unwin, Ltd., 1903.

Russell, J. L. "Time in Christian Thought." In *The Voices of Time,* edited by J. T. Frazer. New York: George Braziller, 1966, pp. 59–77.

Russell, Peter. *The Global Brain.* Los Angeles: J. P. Tarcher, Inc., 1983.

Sagan, Carl. *The Dragons of Eden.* New York: Random House, 1977.

Santillana, Giorgio de. *The Origins of Scientific Thought.* Chicago: University of Chicago Press, 1961.

Schrödinger, Erwin. *What Is Life? & Other Scientific Essays.* Garden City, NY: Doubleday & Co., 1956.

Sessions, Roger. "The Composer and His Message." In *The Intent of the Artist,* edited by Augusto Centeno. Princeton, NJ: Princeton University Press, 1941.

Shapley, Harlow. *Of Stars and Men.* Boston: Beacon Press, 1958, 1964.

Sharnoff, Sylvia Duran. "Lowly Lichens Offer Beauty—and Food, Drugs, and Perfume." *Smithsonian* (April 1984), pp. 135–143.

Shaw, George Bernard. Preface to *Back to Methuselah.* Baltimore: Penguin Books Inc., 1921, 1949.

Simpson, George Gaylord. *The Meaning of Evolution.* New Haven: Yale University Press, 1949.

Sinnott, Edmund. *The Biology of the Spirit.* New York: The Viking Press, 1955.

———. *Matter, Mind, and Man.* New York: Atheneum, 1962.

Sisson, Robert F. "Life Cycle of a Coral." *National Geographic* (June 1973), pp. 780–793.

Smith, Huston. "Let There Be Light." In *Great Religions of the World*. Washington, D.C.: The National Geographic Society, 1971.

———. "The Revolution in Western Thought." *The Saturday Evening Post* (August 26, 1961).

Spencer, Herbert. *Principles of Sociology*. New York: D. Appleton and Company, 1882.

Stace, W. T. "Man Against Darkness." *Atlantic Monthly*, vol. 182, no. 3 (1948), pp. 53–58.

Sussman, Maurice. *Growth and Development*. Englewood Cliffs, NJ: Prentice-Hall, Inc., 2nd edition, 1964.

Teale, Edwin Way. *The Golden Throng*. New York: Dodd, Mead & Company, 1961.

Teilhard de Chardin, Pierre. *The Phenomenon of Man*. English translation by Bernard Wall. New York: Harper & Brothers Publishers, Inc., 1959.

Thomas, Lewis. *Late Night Thoughts on Listening to Mahler's Ninth Symphony*. New York: The Viking Press, 1983.

———. *The Lives of a Cell*. New York: The Viking Press, 1974.

———. *The Medusa and the Snail*. New York: Viking Penguin, 1979. Bantam Books, 1980.

Thomsen, Dietrick. "A Place in the Sun for Fractals." *Science News*, vol. 121 (January 9, 1982), pp. 28, 30.

———. "Making Music—Fractally." *Science News*, vol. 117 (March 22, 1980), pp. 187, 190.

Thomson, William. "Secular Cooling of the Earth." Edinburgh: *Transactions of the Royal Society of Edinburgh*, vol. 23, 1864.

Trefil, James. "Closing in on Creation." *Smithsonian* (May 1983), pp. 32–51.

———. "How the Universe Will End?" *Smithsonian* (June 1983), pp. 73–83.

———. *The Moment of Creation*. New York: Charles Scribner's Sons, 1983.

Van Nostrand Encyclopedia of Science. Princeton, NJ: Van Nostrand, 1958.

Waldrop, M. Mitchell. "A Flower in Virgo." *Science*, vol. 215, no. 4535 (Feb. 19, 1982), pp. 953–955.

Washburn, S. L., and Ruth Moore. *Ape into Man*. Boston: Little, Brown and Company, 1974.

Weinberg, Steven. *The First Three Minutes.* New York: Basic Books, Inc., 1977.

Weisskopf, Victor F. "The Frontiers and Limits of Science." *Daedalus,* Summer 1984, pp. 177–195.

———. *Knowledge and Wonder.* 2nd edition. Cambridge: The MIT Press, 1979.

Wells, H. G., Julian S. Huxley, and G. P. Wells. *The Science of Life.* New York: The Literary Guild, 1934.

White, Edward. "The Chinese Landscape." *Horizon,* vol. 17, no. 4 (Autumn 1975), pp. 87–97.

Whitehead, Alfred North. *Science and the Modern World.* New York: The Macmillan Company, 1925; The New American Library, 1948.

Young, Louise B., editor. *The Evolution of Man.* New York: Oxford University Press, 1970.

———. *Exploring the Universe,* 2nd edition. New York: Oxford University Press, 1971.

———. *The Mystery of Matter.* New York: Oxford University Press, 1965.

Yunis, Jorge J., and Om Prakash. "The Origin of Man: A Chromosomal Pictorial Legacy." *Science,* vol. 215, no. 4539 (March 19, 1982), pp. 1525–1530.

Zervos, Christian. "Conversations with Picasso." Translated by Brewster Ghiselin. *Cahiers d'Art.* Paris: 1935.

ACKNOWLEDGMENTS

The ideas presented in this book represent a synthesis of many thoughts, insights, and observations that have impressed themselves on my consciousness over the last twenty-five or thirty years. These bits and pieces have been reorganized in my mind and combined with perceptions of my own to create a new whole with properties that transcend the characteristics of the parts that compose it. I can no longer identify the origin of every idea, because the thoughts have long ago been taken over and become part of me. But certain very important influences do stand out in my memory—the work of Julian Huxley, Pierre Teilhard de Chardin, Henri Bergson, Loren Eiseley, and perhaps a dozen other writers, philosophers, and scientists. These authors are quoted in the text, and their contributions are gratefully acknowledged.

A number of distinguished contemporary scientists have been generous with their time and expertise in helping me put together this exposition, which draws evidence from many different fields of science. They have read and commented on various pieces of the manuscript, offering valuable suggestions and criticisms. For such very constructive help I am indebted to Jane Overton, Professor of Molecular Genetics and Cell Biology, University of Chicago; Cheves T. Walling, Professor of Chemistry, University of Utah; Robert R. Wilson, the Michael Popin Professor of Physics, Columbia University; S. L. Washburn, University Professor Emeritus of Anthropology, University of California, Berkeley; James E. Lovelock, Visiting Professor, Reading University, England; and John Beckett, Geologist, University of Chicago.

In spite of the expert help and advice I have received, final responsibility for the accuracy of the information and, above all, for the interpretation of the facts is mine alone.

I am also grateful to my editors at Simon and Schuster, who have guided the development of this book from a first rough outline to its present form. They have been a constant source of encouragement and enthusiasm for the thoughts that I have presented here. We have worked well together because we share a belief in the power of a book to preserve and disseminate an idea so it becomes part of our cultural heritage and can bear fruit when the time is ripe.

LBY
Winnetka, Illinois
May 1985

INDEX

PICTURE CREDITS